RENEWABLE ENERGY: RESEARCH, DEVELOPMENT AND POLICIES

UTILITY CUSTOMER-FUNDED ENERGY EFFICIENCY PROGRAMS

PROJECTED SPENDING AND SAVINGS

RENEWABLE ENERGY: RESEARCH, DEVELOPMENT AND POLICIES

Additional books in this series can be found on Nova's website under the Series tab.

Additional E-books in this series can be found on Nova's website under the E-book tab.

RENEWABLE ENERGY: RESEARCH, DEVELOPMENT AND POLICIES

UTILITY CUSTOMER-FUNDED ENERGY EFFICIENCY PROGRAMS

PROJECTED SPENDING AND SAVINGS

TREVOR MASELLI

EDITOR

New York

For permission to use material from this book please contact us:
Telephone 631-231-7269; Fax 631-231-8175
Web Site: http://www.novapublishers.com

NOTICE TO THE READER

Additional color graphics may be available in the e-book version of this book.

Library of Congress Cataloging-in-Publication Data

ISBN: 978-1-62948-939-1

Published by Nova Science Publishers, Inc. † New York

CONTENTS

PREFACE

Electric and natural gas energy efficiency in the United States is pursued through a diverse mix of policies and programmatic efforts, which support and supplement private investments by individuals and businesses. These efforts include federal and state minimum efficiency standards for electric and gas end-use products; state building energy codes; a national efficiency labeling program (ENERGY STAR®); tax credits; and a broad array of largely incentive-based programs for consumers, funded primarily by electric and natural gas utility customers. This book provides information on projected spending and savings of the future of utility customer-funded energy efficiency programs in the United States; and discusses the shifting landscape of ratepayer-funded energy efficiency in the United States.

Chapter 1 - Electric and natural gas energy efficiency in the United States is pursued through a diverse mix of policies and programmatic efforts, which support and supplement private investments by individuals and businesses. These efforts include federal and state minimum efficiency standards for electric and gas end-use products; state building energy codes; a national efficiency labeling program (ENERGY STAR®); tax credits; and a broad array of largely incentive-based programs for consumers, funded primarily by electric and natural gas utility customers. Over the past four decades, policy support and utility customer funding of energy efficiency programs, in particular, has ebbed and flowed. Utilities first launched substantial programs in the wake of the 1973 energy crisis, and those programs grew and matured with the expansion of integrated resource planning and demand-side management during the 1980s and 1990s. Spending on energy efficiency by utilities then declined sharply in many states in the late 1990s, with the restructuring of the electricity industry. However, the western energy crisis of

2000-2001 brought renewed attention to energy efficiency as an important strategy for managing and containing costs for electric utility customers.

Chapter 2 - Over the last two decades, utility ratepayer funding for energy efficiency programs has seen both booms and busts. Currently, roughly 35 states implement some set of ratepayer-funded electric and/or natural gas energy efficiency programs, with a total U.S. budget of $3.1 billion in 2008 (CEE 2008). The top 10 states account for about 80% of this total, each of which planned to spend $100 million or more in 2008. California, the undisputed heavyweight in terms of the absolute magnitude of its spending on energy efficiency, represented one-third of the total U.S. energy efficiency program budget in 2008.

A proliferation of new state-level policies enacted over the past several years suggests that the next decade may see a dramatic and sustained increase in overall funding levels, and a fundamental re-drawing of the energy efficiency map. These new policies include: energy efficiency portfolio or resource standards (EEPS or EERS), requirements that utilities acquire all cost-effective energy efficiency, strengthened integrated resource planning (IRP) or demand-side management (DSM) planning requirements, and regulatory incentive mechanisms to better align utility financial interests with improvements in customer energy efficiency.

In: Utility Customer-Funded Energy … ISBN: 978-1-62948-939-1
Editor: Trevor Maselli

Chapter 1

THE FUTURE OF UTILITY CUSTOMER-FUNDED ENERGY EFFICIENCY PROGRAMS IN THE UNITED STATES: PROJECTED SPENDING AND SAVINGS TO 2025*

Galen L. Barbose, Charles A. Goldman, Ian M. Hoffman and Megan Billingsley

This work was supported by the National Electricity Delivery Division of the U.S. Department of Energy's Office of Electricity Delivery and Energy Reliability under Lawrence Berkeley National Laboratory Contract No. DE-AC02-05CH11231.

Prepared for the Office of Electricity Delivery and Energy Reliability National Electricity Delivery Division U.S. Department of Energy

---

* This is an edited, reformatted and augmented version of Ernest Orlando Lawrence Berkeley National Laboratory, Publication No. LBNL-5803E, dated January 2013.

Abstract

We develop projections of future spending on, and savings from, energy efficiency programs funded by electric and gas utility customers in the United States, under three scenarios through 2025. Our analysis, which updates a previous LBNL study, relies on detailed bottom-up modeling of current state energy efficiency policies, regulatory decisions, and demand-side management and utility resource plans. The three scenarios are intended to represent a range of potential outcomes under the current policy environment (i.e., without considering possible major new policy developments).

Key findings from the analysis are as follows:

- By 2025, spending on electric and gas efficiency programs (excluding load management programs) is projected to double from 2010 levels to $9.5 billion in the medium case, compared to $15.6 billion in the high case and $6.5 billion in the low case.
- Compliance with statewide legislative or regulatory savings or spending targets is the primary driver for the increase in electric program spending through 2025, though a significant share of the increase is also driven by utility DSM planning activity and integrated resource planning.
- Our analysis suggests that electric efficiency program spending may approach a more even geographic distribution over time in terms of absolute dollars spent, with the Northeastern and Western states declining from over 70% of total U.S. spending in 2010 to slightly more than 50% in 2025, and the South and Midwest splitting the remainder roughly evenly.
- Under our medium case scenario, annual incremental savings from customer-funded electric energy efficiency programs increase from 18.4 TWh in 2010 in the U.S. (which is about 0.5% of electric utility retail sales) to 28.8 TWh in 2025 (0.8% of retail sales).
- These savings would offset the majority of load growth in the Energy Information Administration's most recent reference case forecast of retail electricity sales through 2025, given specific assumptions about the extent to which future energy efficiency program savings are captured in that forecast.
- The pathway that customer-funded efficiency programs ultimately take will depend on a series of key challenges and uncertainties associated both with the broader market and policy context and with the implementation and regulatory oversight of the energy efficiency programs themselves.

1. Introduction

Electric and natural gas energy efficiency in the United States is pursued through a diverse mix of policies and programmatic efforts, which support and supplement private investments by individuals and businesses. These efforts include federal and state minimum efficiency standards for electric and gas end-use products; state building energy codes; a national efficiency labeling program (ENERGY STAR®); tax credits; and a broad array of largely incentive-based programs for consumers, funded primarily by electric and natural gas utility customers (Dixon et al. 2010).[1] Over the past four decades, policy support and utility customer funding of energy efficiency programs, in particular, has ebbed and flowed.[2] Utilities first launched substantial programs in the wake of the 1973 energy crisis, and those programs grew and matured with the expansion of integrated resource planning and demand-side management during the 1980s and 1990s (Nadel 1992). Spending on energy efficiency by utilities then declined sharply in many states in the late 1990s, with the restructuring of the electricity industry. However, the western energy crisis of 2000-2001 brought renewed attention to energy efficiency as an important strategy for managing and containing costs for electric utility customers.[3]

Since then, many state regulatory agencies and legislatures have sought to prioritize energy efficiency, in some cases strengthening and supplementing pre-existing policies by requiring comprehensive electric and gas system resource planning, developing funding mechanisms and energy savings targets, and creating business incentives for program administrators to deliver energy efficiency to customers. In some states, regulators have also extended demand-side planning, savings targets, or business incentive mechanisms from the electricity sector to large regulated natural gas utilities.

A variety of organizations and analysts have examined trends in utility customer-funded energy efficiency programs in the United States. These include efforts to document historical and recent trends in spending, savings or both (Nadel 1992; Sciortino et al. 2011; Cooper and Wood 2012; CEE 2012), as well as estimates of the projected impact of individual policies related to utility customer-funded energy efficiency (Nowak et al. 2011) or in particular regions (Hopper et al. 2008). Yet other studies have sought to estimate the potential savings that could be obtained through customer-funded efficiency programs, including an innumerable number of such studies conducted for individual utilities or states, as well as several national studies (EPRI 2009). The present study builds upon the body of existing literature by comprehensively assessing the potential impact of the full suite of policies and market conditions relevant to the future of utility customer-funded energy efficiency programs in the United States, updating an earlier LBNL analysis (Barbose et al. 2009).

Specifically, we project future spending on, and savings from, U.S. electric and gas efficiency programs to 2025 under low, medium, and high scenarios. The projections are based on a detailed, bottom-up review and modeling of all relevant state policies and legislation, regulatory decisions, and utility integrated resource and demand-side management plans. The three scenarios are intended to represent a range of potential outcomes under the *current* policy environment, given uncertainties in policy implementation and in the broader economic and policy environment (e.g., utility business models, the extent to which energy efficiency is a policy priority, and concerns about rate impacts). The three scenarios are not intended to encompass major new federal policy developments, which could naturally expand the range of potential outcomes beyond those modeled here.[4] Scenario definitions and assumptions were also informed by interviews with regional and national energy efficiency experts, program administrators, regulatory staff and other industry stakeholders. Based on the quantitative analysis of projected spending and savings under varying policy implementation scenarios, we identify and discuss the broader themes and issues that will influence which of thc potential projections are most likely to transpire.

The study has relevance to a broad range of audiences: utilities and other entities responsible for administering customer-funded efficiency programs and the state regulatory agencies responsible for overseeing their implementation; policymakers, planners, and industry analysts seeking to understand the potential impact of these programs on the broader electricity market or their implications for other policies; and the energy services industry seeking to understand market trends and opportunities. While this study focuses on the United States, the analysis also has relevance to policymakers abroad where energy and environmental policies may require the development of specific long-term energy savings goals and/or funding mechanisms for voluntary incentive-based programs, such as those that are prevalent in the United States. For example, the set of potential trajectories of U.S. efforts potentially offers a window on the prospects and issues raised by the 2012 European Union Energy Efficiency Directive (Directive 2012/27/EU), in which the European Parliament and Council committed member states to adopting efficiency targets and submitting implementation plans consistent with a EUwide target of saving 20% of the projected primary energy consumption in 2020 (Boonekamp 2011).

The remainder of the paper is organized as follows. Section 2 provides an overview of the key policy drivers that influence future efficiency program spending and savings, and summarizes current trends in spending on energy efficiency programs across states. Our modeling approach for capturing policy and market influences on future spending and savings for electric and gas efficiency programs is described in Section 3. The results of our analysis are presented in Section 4. In Section 5, we identify key challenges and discuss significant uncertainties in market and policy drivers that may influence the path forward for customer-funded efficiency programs.

2. Recent Policy Developments and Trends

Over the past decade, an increasing number of states have adopted policies that encourage or require utility customer-funded energy efficiency programs. In this section, we summarize recent trends in the development of these policies and the current and historical spending levels across states.

2.1. Policy Drivers for Utility Customer-Funded Energy Efficiency Programs

In the utility sector, policies that drive investment in energy efficiency include: system benefit charges; energy efficiency resource standards; renewable portfolio standards under which energy efficiency is a qualifying resource; requirements that utilities obtain "all cost-effective energy efficiency" resources; long-term integrated resource planning requirements; and multiyear demand-side management planning requirements (see Table 1). Naturally, the scope and level of aggressiveness of each type of policy can vary substantially across states, and many states have adopted multiple policies in tandem.

A number of these policy drivers are relatively recent, most notably energy efficiency resource standards (EERS), which have thus far been adopted in 15 states and require utilities to achieve minimum energy efficiency savings targets over a lengthy period of time.[5] Similarly, several others states have adopted broader renewable portfolio standards (RPS) or alternative energy standards under which energy efficiency is a qualifying resource. Many of these EERS policies and RPS policies with energy efficiency allowances have been enacted in states that previously had not aggressively pursued customer-funded energy efficiency and have therefore required rapid development of the regulatory and administrative structures necessary to implement and oversee sizable energy efficiency program portfolios. Another recent policy development in a number of states, all of which have offered large-scale energy efficiency programs for more than a decade, is the development of statutory or regulatory requirements that utilities acquire "all cost effective" energy efficiency. In these states, program administrators or regulatory staff may then conduct studies that estimate the long-term, cost-effective savings potential and then propose annual or multi-year savings targets and budgets in order to capture this potential over a defined time period. Other facets of the energy efficiency policy landscape are less recent. System benefit charges (SBC), which exist in 14 states and were typically established more than a decade ago as part of larger electric industry restructuring processes, serve to set an approximate floor on energy efficiency program spending via a non-bypassable surcharge on customers' utility bills. Integrated resource planning (IRP) also exists in many states, whereby utilities are required to plan for the long-term needs of their customers by considering and assessing a broad range of resource options, including energy efficiency resources. Depending upon the manner and extent to which utilities are required to assess energy efficiency options, the IRP may culminate in a 10- to 20-year plan with specified levels of energy efficiency resource acquisition. Finally, utilities in many states are required to regularly submit a demand-side management (DSM) plan to their state regulator, proposing a specific portfolio of programs that meet cost-effectiveness guidelines and other policy objectives, typically on a one- to three-year cycle. Although IRP and DSM planning have both been utilized for more than 20 years, their application has expanded somewhat in recent years as a result of policy spill-over or cross-border effects from other states within a given region. For example, Arkansas regulators developed a step-by-step energy efficiency program development template that has been cited as a policy influence in other southern states (e.g., Mississippi and Alabama).[6] Multi-state utilities also are developing territory-wide efficiency programs designed to meet one state's mandates, in effect carrying that state's energy saving policies *de facto* into neighboring states.[7] Lastly, the move by the Tennessee Valley Authority (TVA)[8] to set savings targets through its IRP, and to offer programs and encourage its member distributors to offer programs, is expected to spread the pursuit of energy efficiency across the seven states where it provides wholesale power.

Table 1. Policy Drivers for Customer-Funded Energy Efficiency Program Activity

Key Policy Drivers for Energy Efficiency Spending and Savings	Applicable to Electric Efficiency Programs	Applicable to Natural GasEfficiency Programs
Energy Efficiency Resource Standard (EERS)	AZ, CA, CO, HI, IL, IN, MD, MI, MN, MO, NM, NY, OH, PA, TX	CA, CO, MI, MN, NY, IL
Energy efficiency eligibility under state RPS	HI, MI, NC, NV,OH	
Statutory requirement that utilities acquire all cost-effective energy efficiency	CA, CT, MA, RI, VT, WA	CA, CT, MA, RI, VT, WA
Systems benefit charges	CA[9], CT, DC, MA, ME, MT, NH, NJ, NY, OH, OR, RI, VT, WI	CA, DC, ME, MT, NJ, NY, RI,WI
Integrated resource planning	34 States (primarily in the West and Southeast) and TVA	17 States (primarily in the West and Northeast)
Demand Side Management plan or multi-year energy efficiency budget	28 States	21 States (primarily in the Northeast and Midwest)

In addition to the energy efficiency policy drivers summarized in Table 1, other broad market and policy dynamics may also play a critical role in shaping the trajectory of future spending and savings from customer-funded energy efficiency programs. We discuss several of these factors in Section 5.1, including the timing and pace of the economic recovery, the long-term trend in natural gas prices, the stringency of future federal and state minimum efficiency standards for appliances and building codes, and the outcome of federal air emissions regulations.

2.2. Current and Historical Spending on Utility Customer-Funded Energy Efficiency Programs

Over the latter half of the past decade, spending on electric and gas utility customer-funded energy efficiency programs (excluding load management)[10] more than doubled, from roughly \$2 billion in 2006 to \$4.8 billion in 2010, consisting of \$3.9 billion for electric energy efficiency programs and \$0.8

billion for natural gas programs (CEE 2008; CEE 2012). Approved budgets for 2011 – which may diverge from actual expenditures – were significantly higher than 2010 spending, totaling $6.7 billion, consisting of roughly $5.6 billion for electric efficiency programs and $1.2 billion for gas efficiency programs (CEE 2012). With the steady increase in spending on utility customer-funded efficiency programs in recent years, relative spending as a percentage of utility revenues has also risen, with electric program expenditures in 2010 equivalent to roughly 1.1% of total U.S. electric utility revenues in that year, while gas program expenditures were equivalent to roughly 0.7% of total U.S. gas distribution utility revenues. The geographical distribution of both electric and gas spending has spread over time, as numerous states with recently adopted policies have ramped up their efforts. That said, total spending on utility customer-funded energy efficiency programs, in absolute dollar terms, still remains highly concentrated within a relatively small number of states (see Table 2).[11]

In particular, the majority of funding for electric efficiency programs is concentrated in California, the Pacific Northwest (OR, WA), and the Northeast (MA, NJ, NY, CT), all states with a long history of commitment to energy efficiency. Other states, many located in the Midwest (e.g., OH, PA, IL, IN, and MI), are in the process of ramping up program spending, often driven by long-term electricity savings targets. The top 10 states, in terms of absolute dollar expenditures account for about 70% of 2010 spending on electric energy efficiency programs. Program administrators in the leading states with the highest per capita energy efficiency spending typically offer a comprehensive portfolio of programs tailored to residential, commercial, and industrial customers that utilize a variety of designs and intervention strategies (e.g., technical assistance to end users and trade allies, incentives to customers to buy down the cost of high-efficiency equipment, and incentives to upstream manufacturers and retailers to stock and distribute high-efficiency products).

Gas efficiency programs are less widespread than electric programs, and thus funding is even more highly concentrated in a small number of states, where the top-10 states account for almost 80% of the national budget for gas efficiency programs. Specifically, gas efficiency spending is concentrated in about a dozen states in various regions: NY, MA, and NJ in the Northeast; IL, MI, IA, MN and WI in the Midwest; and CA, OR and UT in the West. Most southern utilities have modest retail gas sales or function largely as distribution entities that convey “transportation gas”,[12] and they consequently spend little on gas efficiency programs. Nationally, gas efficiency program budgets are dominated by residential and low income programs, together comprising 68% of total program expenditures in 2010 (CEE 2012), which is quite different from the program mix for electric efficiency programs.[13]

Table 2. 2010 Expenditures for Utility Customer-Funded Energy Efficiency (EE) Programs

Rank	Combined Electric & Gas EE Spending ($M)		Electric EE Spending ($M)		Gas EE Spending ($M)	
1	CA	$1,139	CA	$938	CA	$201
2	NY	$521	NY	$482	NJ	$126
3	NJ	$317	MA	$245	MA	$72
4	MA	$317	WA	$218	MI	$41
5	WA	$247	NJ	$191	IA	$40
6	FL	$176	FL	$165	NY	$39
7	OR	$158	OR	$135	MN	$36
8	MN	$144	TX	$114	UT	$36
9	CT	$119	CT	$108	OH	$32
10	MI	$116	MN	$107	WA	$29
Top 10 States	*$M*	$3,255		$2,702		$653
	% of U.S.	68%		68%		78%
Remainder of U.S.	*$M*	$1,531		$1,246		$185
	% of U.S.	32%		32%		22%
Total U.S.	*$M*	$4,786		$3,948		$838

Source: CEE (2012), with several modifications.

3. Analytical Approach

We developed low, medium, and high case projections of electric and natural gas efficiency program spending to 2025, as well as accompanying projections of electric program energy savings.[14] These projections are based on a state-by-state review of current policies, regulatory decisions, utility IRPs and DSM plans, and other key regulatory and planning documents, further supported through interviews with state PUC and utility staff and regional energy efficiency experts. The projections are intended to represent alternative pathways for the future evolution of energy efficiency programs, given the *current* set of policies in place and the larger market and policy environment in which programs operate. As explained further below, we took different approaches to developing projections for electric and gas energy efficiency program spending. These methodological differences reflect both that enabling efficiency policies are more prevalent among electric utilities compared to gas utilities, and that the level of development and experience with administering electric efficiency programs is much greater than for gas programs.

3.1. Electric Energy Efficiency Program Spending and Savings Projections

The projections of electric program spending and savings are based primarily on state-specific assumptions about how effectively and aggressively current energy efficiency policies are implemented and about the impact of broader market conditions. The scenario assumptions are summarized by census region in Table 3. The projections for these states typically begin with assumptions about either future spending or savings (depending on the state and scenario), and then future spending or savings are derived from the other based on assumptions about the cost of savings. For a group of seven "uncommitted" states that currently have little efficiency program activity and no established policy framework, we instead employ a standardized approach by which spending increases above current levels by a stipulated amount under each scenario, also described in Table 3.[15] Additional methodological details, including state-by-state descriptions of scenario definitions, are provided in Appendix A.

Although the scenario definitions were developed on a state-by-state basis, with consideration of the specific policy and market context in that state, the low, medium, and high scenarios can be characterized in broad terms. At a conceptual level, the low scenario represents a less prominent role for energy efficiency as a resource in many states, as program spending remains at current levels or increases very modestly (or decreases in a few states) in subsequent years. The medium scenario reflects a future in which states that historically have been leaders in energy efficiency continue down that path and in some cases expand the role of energy efficiency as a resource, while other states are fairly successful in ramping up their energy efficiency programs to meet legislative saving targets. Note that in the medium scenario, our estimates of future savings account for constraints that may limit the ability of program administrators to achieve savings targets – e.g., ability for energy efficiency services infrastructure to ramp up quickly in early years and rate or spending caps that limit program spending increases in later years. The high scenario reflects a future in which many states establish a very prominent role for energy efficiency as a resource: states with EERS statutes are assumed to meet savings targets (and overcome constraints), states in each region are inclined to follow the example (and goals) established by leading states in that region, and those states that are currently "uncommitted" are assumed to adopt policies that lead to savings in 2025 of roughly the national average savings targets achieved by utilities currently.

Table 3. Scenario Assumptions for Electric Energy Efficiency Projections

Region	Scenario	Representative Assumptions
South	Low	TX IOUs meet minimum EERS targets. TVA savings based on 2010 IRP "Baseline Portfolio." NC IOUs achieve only as much savings as can be applied towards their RPS. Utilities maintain spending at the level in the last year of recent DSM plans (in terms of percentage of revenues).
	Medium	TX IOUs maintain savings at current levels (0.2% of sales), exceeding EERS targets. FL IOUs ramp up savings to 0.3% of sales. IRP savings targets are achieved (TVA, KY, GA). Otherwisesimilar to low case.
	High	TX utilities ramp up savings to 0.3% of sales, offsetting 50% of demand growth. FL IOUs ramp up to long-term savings targets established by regulators (0.48% of sales in 2019) and to 0.75% by 2025. TVA and IOUs in GA, NC, and SC ramp up savings to roughly 1% of retail sales. MD utilities meet state EERS goals.
Midwest	Low	IL, IN, and OH utilities fall short of EERS targets (e.g., due to cost caps or opt-out), but MI and MN utilities fully meet their more-modest EERS targets. IA utilities maintain current spending levels, and WI spending is equal to current legislative cap of 1.2% of revenues.
	Medium	EERS targets are achieved in most cases, one exception being IL, where cost cap is eased but not to the extent required to meet ultimate targets. IA spending continues at the level in the last year of the most recent DSM plans. WI spending rebounds to 1.7% of revenues, half way between current legislative cap and historical peak.

Table 3. (Continued)

Region	Scenario	Representative Assumptions
	High	All states reach savings of roughly 1.5% to 2% of retail sales, meeting or exceeding EERS targets by varying degrees.
West	Low	CA IOU savings are based on 90% of market potential, as estimated in Navigant(2012), which decline from current levels. AZ and NM utilities achieve EERS targets. Many utilities in the Northwest achieve savings equal to 60% of NPCC's 6thPower Plan conservation targets, with remainder achieved through codes and standards. Utilities in other states achieve savings based on most recent IRP or maintain constant savings based on the final year of their most recent DSM program plan, whichever is less.
	Medium	CA IOU savings are based on 110% of market potentialin Navigant(2012). CO utilities achieve EERS targets. In the Northwest, utility savings equal 75% of NPCC conservation targets, with remainder achieved through codes/standards. Utilities in most other states achieve savings based on most recent IRP or maintain constant savings based on the last year of their most recent DSM program plan, whichever is greater.
	High	CA IOU savings are based on 130% of market potentialin Navigant (2012). AZ IOUs meet EERS targets without reliance on retroactive credit for historical programs, and SRP achieves similar savings levels. CO is same as medium case. In the Northwest, utility savings equal 85%of NPCC conservation targets, with remainder achieved through codes/standards. Utilities in many other states achieve savings of roughly 1.5% of retail sales.
Northeast	Low	Spending levels in most states remain flat at roughly the statutory minimum (constituting a decline from current spending in some states) and/or continue at current funding levels. In NJ, spending declines by more than 50% from current levels, as reliance shifts to revolving loan funds, with program spending equal to roughly 70% of the levels specified in the recent RFP for program administration.

Region	Scenario	Representative Assumptions
	Medium	Spending in most states, as a percentage of revenues, remains flat at the level in the final year of the most recent energy efficiency program plan. NY meets its EERS target for 2015, but spending thereafter reverts to the 2010-2015 average. PA spending rises to current cap. NJ programs shift to revolving loan funds, but spending levels declines less severely than Low Case to reflect amore successful transition to financing model.
	High	New England IOUs capture all cost-effective energy efficiency, up to a stipulated spending cap (10% of revenues for MA, RI, and VT; 6.5% for CT). NY meets its EERS target for 2015, and spending thereafter continues at 2015 levels. Savings in other states rises to 1-2% of retail sales.
Uncommitted	Low	Spendingincreases to 0.3% of revenues above current levels
	Medium	Spending increases to 0.5% of revenues above current levels
	High	Spending increases to 0.8% of revenues above current levels

3.2. Gas Energy Efficiency Program Spending Projections

For the purpose of developing projections of utility customer funding of gas efficiency programs, we first grouped states into three categories: Tier I consists of the 13 states that comprise more than 80% of current national funding for gas efficiency programs, Tier II consists of another 15 states where 2010 spending on gas efficiency programs exceeded $0.50 per capita, and Tier III consists of the remaining 23 states. (See Technical Appendix A for the set of states included within each tier).

The process for developing scenario definitions for each state differed according to its tier (see Table 4). For Tier I states, gas efficiency program spending projections are based on state-specific policies, gas DSM program plans, and regulatory decisions that set savings targets for gas utilities, and were further informed by interviews with program administrators, regulators and other experts in the field. For most Tier I states, the low and medium case spending projections track the most recent multi-year gas DSM program plans to their terminal year (typically 2012 to 2014). In the low case, we assume that spending on residential gas efficiency programs in most Tier 1 states will decline to 25% of the level in the terminal year of the most recent DSM plan, while spending on commercial and industrial (C&I) programs will decline to roughly 80% of the level in the terminal year of the DSM plan. This decline in spending is due to the combination of sustained low natural gas prices, which reduce the cost effectiveness of gas efficiency programs, and tightening federal minimum efficiency standards for gas furnaces, which reduce the savings for voluntary programs – both of which are discussed further in Section 5. In the medium case, we assume a more modest drop-off in residential program spending, typically to 50% of the level from the terminal year of the most recent gas DSM Plan, but that C&I program spending increases slightly, as program managers shift budgets towards markets with greater savings opportunities. In both the low and medium scenario, we assume that spending on gas low-income programs remains constant at the level from the last year of the DSM plan, as these programs meet broader policy objectives (e.g. equity, reductions in bill arrearages) and therefore are less susceptible to the dynamics putting downward pressure on gas program spending for the other sectors. Finally, in the high case, we assume that many Tier I states achieve gas savings levels on par with the gas EERS targets recently adopted in several states (i.e., generally 1.0-1.5% of total gas distribution utility retail sales).

The 15 Tier II states have relatively aggressive spending levels on a per capita basis, but small populations and therefore small spending levels in absolute terms. Thus, for simplicity, the spending projections for these states were developed based on regional benchmark trajectories that were developed from the projections for Tier I states in the corresponding region. These regional benchmark trajectories were developed by averaging the change in spending as a percentage of gas distribution utility revenues per year by region for the Tier I states in each census region. Those growth curves were then applied to the 2010 spending for each Tier II state. As an example, in the medium case, spending for the three Tier I states in the Northeast (MA, NY, and NJ) is projected to increase by, on average, 0.6% of revenues; thus, the same 0.6% increase in spending as a percent of revenues was stipulated for the Tier II northeastern states in the medium case. For further details, please refer to Technical Appendix A.

For the remaining 23 Tier III states that currently have little or no customer-funded gas program activity, we assumed that future gas efficiency spending will, in the low case, remain at 2010 levels in absolute nominal dollar terms (thus declining as a percentage of gas distribution utility revenues, as revenues grow). In the medium case, we assume that program administrators maintain gas efficiency spending at their present level, in terms of the percentage of utility revenues. The high case posits that program administrators will increase program spending to approximately 0.25% of revenues above 2010 levels by 2025.

Table 4. Scenario Assumptions for Gas Energy Efficiency Projections

Category	Scenario	Representative Assumptions (specific assumptions vary by state)
Tier I States	Low	Assume new furnace equipment standards and moderate gas prices cause a reduction inresidential program spending to 25% of the level from the most recent gas DSM plan, and to 80% for commercial & industrial programs. No change in low-income program spending.
	Medium	Assume new furnace equipment standards and moderate gas prices cause a reduction inresidential program spending to 50% of the level from the most recent gas DSM plan, but a slight increase in C&I program spending. No change in low-income program spending.
	High	States reach stipulated benchmarks for gas program savings ranging from 1.0% to 1.5% of total gas distribution utility sales.
Tier II States	All	Regional benchmark (average) based on low, medium and high scenarios in Tier I
Tier III States	Low	Spending remains at 2010 levelsin absolute nominal dollar terms
	Medium	Spending remains at 2010 levels as a percentage of gas distribution utility revenues
	High	Spending percentage increases above current levels by 0.25% of gas distribution utility revenues

4. Results

In this section, we present our projections of spending on utility customer-funded energy efficiency programs through 2025. We first present total projected spending for electric and gas efficiency programs, combined, before turning to the projections for each fuel individually. We also present projections of electric energy savings associated with the three spending trajectories for electric efficiency programs and consider the potential significance of these savings projections in relation to current expectations about future load growth in the electric sector.[16] The results presented throughout this section focus primarily on national and regional trends; the corresponding state-level projections are provided in Technical Appendix B.

4.1. Combined Electric and Gas Energy Efficiency Program Spending Projections

Total spending on electric and gas energy efficiency programs is expected to increase in all scenarios across the study period. By 2025, we project that total electric and gas efficiency program spending, in nominal dollars, will rise from \$4.8 billion in 2010 to \$6.5 billion in the low case, \$9.5 billion in the medium case, and \$15.6 billion in the high case (see Figure 1). These projections correspond to compound growth rates of approximately 2% per year (low case), 5% per year (medium case), and 8% per year (high case). Although the projected increase in spending in both the medium and high cases is sizable in absolute dollar terms, the associated growth rates in all cases are substantially lower than that witnessed over the past half-decade, when total electric and gas efficiency program rapidly accelerated at an average rate of 26% per year from 2006 to 2010 (Eldridge et al. 2008, CEE 2012). In the decade preceding this recent and rapid expansion of energy efficiency program activity, however, electric program spending grew by less than 5% per year from 1997 to 2006, which is on par with the projected growth in spending under the medium case.

As discussed further in the following sections, projected growth rates for electric efficiency program spending are somewhat higher than for gas program spending in both the low and medium cases, with projected electric program spending growth of 2.3% and 4.9% per year in the low and medium cases, versus less than 1.1% and 3.8% per year for gas programs. In the high case, however, gas efficiency spending grows faster than electric spending (9.7% vs. 7.8%). These differing trends reflect, in large part, the broader base of underlying policy support for, and historical experience with, electric efficiency programs, leading to stronger growth in the low and medium cases for electric programs, while leaving a large upside potential for growth in gas program spending under the high-case conditions.

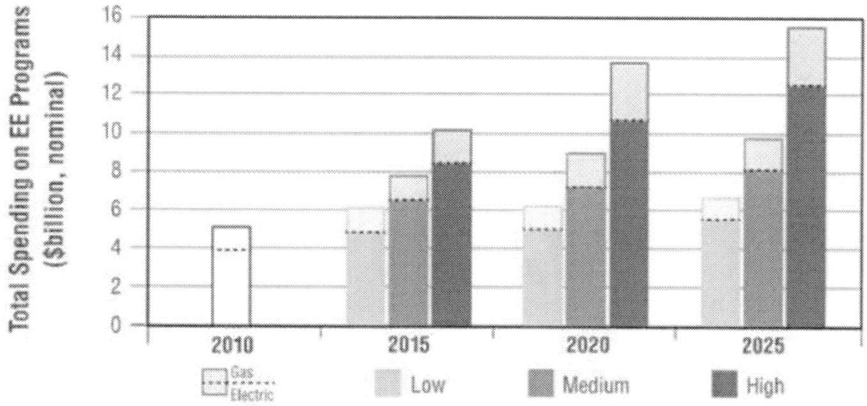

Source: 2010 spending based on CEE (2012).

Figure 1. Projected Electric and Gas Energy Efficiency Program Spending.

4.2. Electric Energy Efficiency Program Spending Projections

Spending on electric utility customer-funded energy efficiency programs is expected to increase, in nominal dollar terms, across all scenarios (see Table 5). Relative to 2010 spending of $3.9 billion (1.1% of total electric utility retail revenues), spending is expected to more than double to $8.1 billion by 2025 in the medium case (1.7% of revenues). In comparison, spending in the low case is projected to increase more slowly, reaching $5.5 billion by 2025 (1.1% of revenues). As described in Section 3, this slower pace of spending growth reflects a future scenario in which regulators and administrators "stay the course" at current funding levels, and many states with aggressive savings targets fall short of those goals. In the high case, spending more than triples from 2010 levels, reaching $12.2 billion (2.7% of revenues), due to the impact of "all cost effective efficiency" policies in leading states, successful achievement of EERS targets, and an increase in program savings in a number of states to the levels projected for regional peers.

Importantly, the projected growth in electric program spending across all cases does not occur smoothly over the forecast period, but rather is "front-loaded", with much faster growth projected through 2015 (Table 5). In the medium case, for example, spending grows by 11% per year through 2015 but by only 2% per year from 2020 to 2025. This dynamic is partly due to the fact that, in many states, recent multi-year DSM plans entail significant spending increases over the next several years, but no longer-term targets or resource planning process currently exists to guide program activity beyond the time horizon of the DSM plan. The front-loaded spending projections also reflect the trajectory of EERS schedules, which typically reach their terminal targets by 2020 or sooner. From 2020 onward, we assume that spending growth in many states tapers off and grows roughly in proportion with projected revenues, reflecting both a lack of strong policy drivers for continued spending growth after 2020, as well as the assumption that savings potential within the 2020-2025 period will be diminished due to the success of programs implemented over the prior decade and tightening federal efficiency standards.

Table 5. Projected Electric Energy Efficiency Spendin

Scenario	Projected Spending ($B, nominal)			Projected Spending (% of Revenues)			Average Annual Spending Growth		
	2015	2020	2025	2015	2020	2025	2010-2015	2015-2020	2020-2025
Low	4.8	5.2	5.5	1.2%	1.2%	1.1%	4%	2%	1%
Medium	6.5	7.4	8.1	1.7%	1.8%	1.7%	11%	3%	2%
High	8.3	10.8	12.2	2.2%	2.6%	2.7%	16%	5%	3%

Not surprisingly, total U.S. electric program spending across all scenarios are driven, in large measure, by EERS policies, energy efficiency eligibility under RPS policies and legislative mandates requiring utilities to acquire all cost-effective energy efficiency. In the medium case, for example, the 15 states with an electric EERS, plus the additional five states with legislative "all cost-effective energy efficiency" mandates (and no associated EERS) and the two states that qualify energy efficiency as an eligible resource under a renewable portfolio standard (again, without an associated EERS) together account for 72% of the total projected electric efficiency program spending in the U.S. in 2025 (see Figure 2). The remaining spending is associated primarily with the additional 18 states that rely primarily on DSM planning and/or IRP (without an associated EERS or "all cost-effective energy efficiency" mandate) to establish their electric efficiency budgets and targets, together comprising 28% of total projected spending on electric efficiency programs.

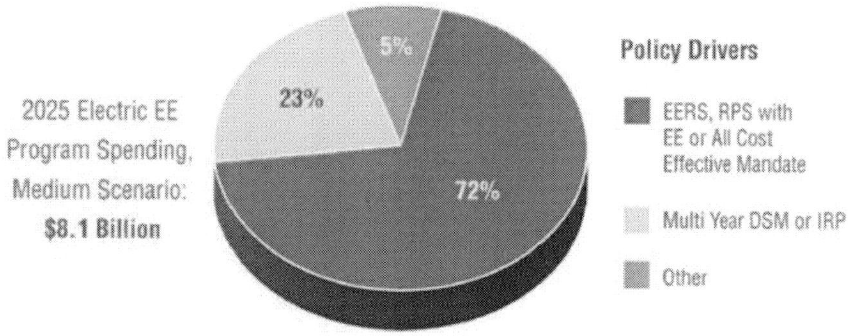

Figure 2. Policy Drivers for Projected Electric Energy Efficiency Spending in the Medium Case (2025).

Projected trends in total U.S. spending are, to some extent, an overlay of distinct quasi-regional trends (see Figure 3). In the medium scenario, overall growth of national efficiency program spending is driven chiefly by projected growth in the Midwest and South, which together represent 70% of projected total U.S. electric program spending growth over the 2010-2025 period. In the Midwest, spending growth is associated with a contingent of populous states (IL, IN, MI, OH) that are currently ramping up to meet statutory EERS targets, while in the South, increases in efficiency program spending are associated with a collection of relatively modest EERS policies and nascent IRP/DSM planning processes in states with a large base of energy consumption (TX, FL, NC, MD, KY). The same underlying policy drivers propel spending growth in these two regions in the low and high scenarios as well, though to differing degrees.

In the West and Northeast – the traditional bastions of energy efficiency activity – electric program spending is also projected to increase in the medium case, though by lesser amounts than the other two regions, reflecting the more mature state of those markets. In the Northeast, efficiency program spending is projected to increase under all three scenarios, where differences in spending levels between the medium and high cases are largely driven by assumptions about how utility program administrators and state regulators translate statutes requiring acquisition of all cost-effective efficiency into multi-year savings goals. For the West, the regional trends are dominated by California, where electric program spending in both the medium and low cases is projected to decline over the long term, as saturation within key end-use markets occurs and as the state leans more heavily on other energy efficiency policies (Navigant 2012). In the medium case, those declines are offset by spending growth in other western states, leading to net spending growth for the region as a whole, while in the low case, total electric program spending in the West is projected to decline slightly.

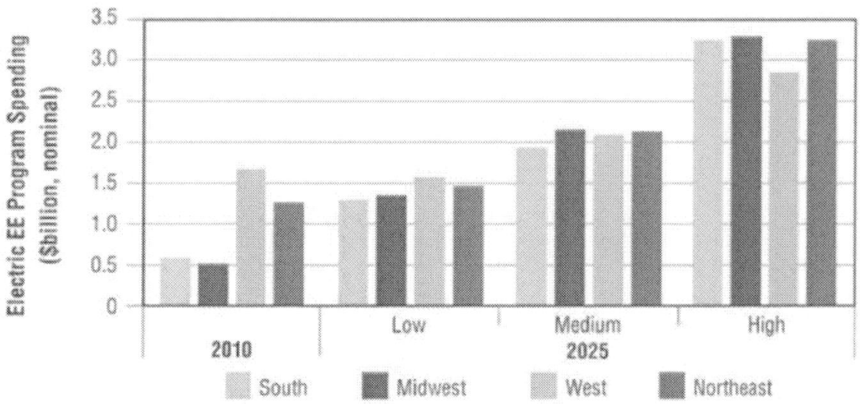

Figure 3. Projected Electric Energy Efficiency Spending by Census Region.

The differing regional trends imply a continued shifting of the energy efficiency map over the coming decade and beyond (see Figure 4). While states in the West and Northeast accounted for more than 70% of efficiency program spending in 2010, that percentage declines to just over 50% by 2025 in the medium case, with the South and Midwest splitting the remaining spending about evenly.

Notwithstanding the greater regional balance in *absolute* dollar spending on electric efficiency programs, the South is still projected to lag well behind other regions in terms of *relative* spending levels as a percentage of electric utility revenues. As shown in Figure 5, spending as a percentage of revenues in the medium case is projected to rise from 1.8% and 2.8% in the Northeast over the 2010 to 2025 timeframe, and decline slightly from 2.4% to 2.1% in the West.

In the Midwest, efficiency spending is expected to increase quite dramatically (from 0.7% to 2.2% of revenues). However, in the South, while spending as a percentage of total electric utility revenues rises from 0.4% of revenues in 2010 to 0.9% in 2025, this is one-third to one-half the spending levels projected in the other three regions.

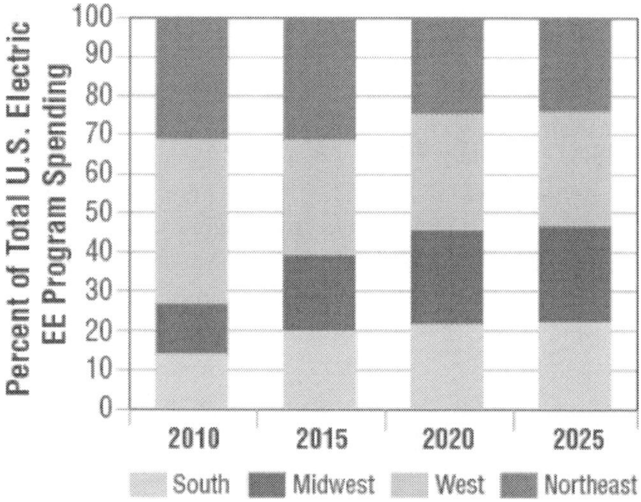

Figure 4. Regional Distribution of Electric Energy Efficiency Program Spending (Medium Case).

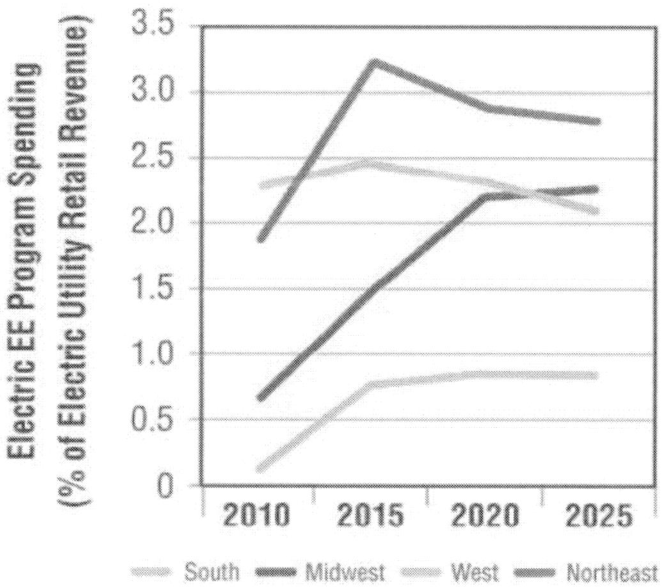

Figure 5. Electric Energy Efficiency Program Spending as a Percent of Utility Revenues (Medium Case).

4.3. Electric Energy Efficiency Program Savings Projections

In 2010, electric energy efficiency programs in the U.S. achieved incremental energy savings of 18.4 TWh, equivalent to 0.49 % of electric utility retail sales nationally (Foster et al. 2012).[17] In comparison, leading states, where program administrators typically have a decade or more of experience in delivering energy efficiency programs, have achieved annual savings of more than 1.0% of retail sales (e.g., CA, CT, MA, OR, VT, NV, HI, RI, and MN), and a sizeable contingent of other states has consistently achieved savings in excess of 0.50% of retail sales.

As explained previously in Section 3 (and in greater detail in Appendix A), the electric efficiency program *spending* projections are linked to a corresponding set of *savings* projections (see Table 6 and Figure 6), where in some cases savings estimates are derived from spending, and in other cases vice-versa.[18] In the medium case, incremental annual energy savings from electric efficiency programs are projected to increase to 28.8 TWh and 0.76% of retail sales in 2025. This represents roughly a 50% increase over the impact of electric efficiency programs in 2010. As was the case for the spending projection, much of the projected increase in annual incremental savings is concentrated in the initial years of the forecast period, as the projection follows the trajectory of the most recent batch of utility energy efficiency plans (which typically terminate in the 2012-2014 period) and EERS targets (which typically reach their final percentage targets by 2020 or sooner).[19] In the low case, incremental annual savings rise moderately by 2015 before largely flattening out over the remainder of the forecast period, reaching 20.6 TWh or 0.53% of retail sales by 2025. In the high case, annual incremental savings rise to 41.6 TWh by 2025, more than double the level achieved in 2010, equivalent to 1.13% of total electric utility retail sales. Thus, in effect, the high case represents a scenario in which the national average savings rise to the level currently being attained by the top tier of states. In both the medium case and the high case, savings levels nationally are within the bounds of most studies of "achievable" energy efficiency potential. This suggests, among other things, that the level of savings projected in these two cases could potentially be reached through accelerated deployment of current technologies, without significant reliance on new efficiency technologies.

Table 6. Projected Incremental Annual Electricity Savings from Utility Customer-Funded Programs (TWh)

Scenario	2010	2015	2020	2025
Low	18.4	20.4	21.1	20.6
Medium		26.6	28.6	28.8
High		33.1	39.8	41.6

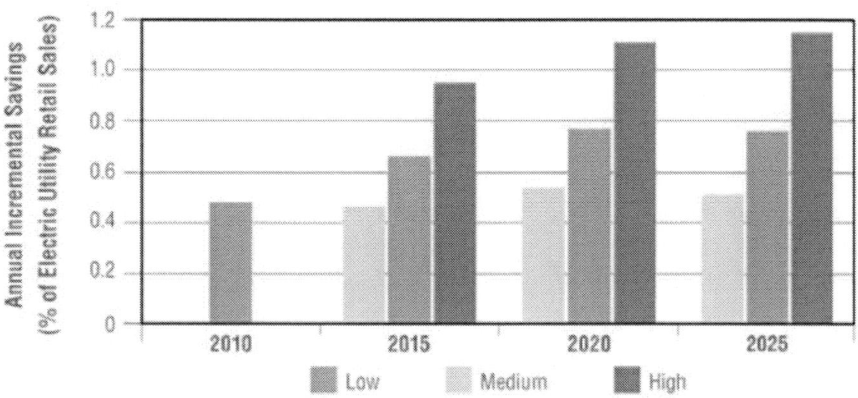

Figure 6. Projected Electricity Savings from Utility Customer-Funded Efficiency Programs.

To place these savings projections in perspective, the Energy Information Administration (EIA)'s most recent reference case forecast (EIA 2012) projects that total U.S. retail electricity sales will grow at a compound annual growth rate (CAGR) of 0.58% over the 2010 to 2025 period, which is substantially lower than the average U.S. load growth of 1.6% per year over the past two decades. The EIA's modeling framework (the National Energy Modeling System, or NEMS) does not explicitly account for the impacts of future utility customer-supported efficiency programs; however, the electricity sales forecast generated by NEMS implicitly assumes that historical trends in utility customer-funded efficiency programs will continue over the forecast period. For the period 2000 to 2010, we estimate that utility customer-funded energy efficiency programs nationally achieved incremental savings of roughly 0.18% per year, on average.[20] Thus, if one were to assume that the EIA reference case forecast implicitly assumes that savings from customer-funded electric efficiency programs continue to accrue at this historic rate, then a hypothetical reference case forecast with no future customer-funded energy efficiency activity would correspond to a CAGR of 0.76% (i.e., 0.58% plus 0.18%).

Our medium case projection corresponds to average annual incremental electricity savings of 0.72% of retail electric sales per year between 2010 and 2025. This, in turn, implies that if electric utility customer-funded efficiency programs achieve savings at the level projected under our medium case, they would reduce growth in U.S. retail electricity sales to just 0.04% per year through 2025 (i.e., 0.76% annual growth with no future efficiency program activity minus projected annual incremental savings of 0.72% of retail sales per year under the median case), offsetting almost all projected load growth under EIA's 2012 reference case forecast.[21] Following the same logic, our low case and high case savings projections would offset roughly 70% and 120% of load growth, respectively, yielding average annual growth rates for retail electricity sales of 0.21% and -0.18% from 2010 to 2025. To be sure, these benchmarks should be considered no more than a "back-of-the-envelope" estimate of the impact of projected customer-funded efficiency program savings on load growth in the United States. Nevertheless, they suggest that rising savings levels, in combination with modest underlying drivers for load growth, could potentially lead to flat, or even negative, load growth over the next 10 to 15 years.

4.4. Gas Energy Efficiency Program Spending Projections

Our analysis suggests a very different set of trajectories for gas efficiency programs compared to electric efficiency programs (see Figure 7 and Figure 8). While the low and medium scenarios both show gas efficiency spending increasing from 2010 to 2015, associated primarily with increases that have already been planned or approved in recent multi-year gas DSM plans, we currently see little evidence to expect significant further spending growth at a national level beyond 2015. Thus, in the low case, spending on gas efficiency programs recedes from its elevated level in 2015 to below $1 billion in 2025 (0.5% of revenues), which is slightly higher than 2010 spending in absolute nominal dollar terms but lower as a percentage of gas distribution utility revenues. In the medium case, spending remains roughly flat at projected 2015 levels, reaching almost $1.5 billion in 2025, equivalent to 0.8% of revenues, a slight increase over the 2010 level. As discussed in Section 3, the low and medium case projections are driven largely by scheduled increases in federal minimum efficiency standards for furnaces, with differing assumptions between the low and medium cases about the extent of the impact on residential gas efficiency spending and the degree to which declines in residential program spending may be offset by increased spending on programs that target commercial/industrial customers. In the high case, however, where gas program savings in the leading states are assumed rise to levels on par with current leading states for electric efficiency, spending on gas programs roughly triples from 2010 levels, reaching $3.3 billion in 2025 (1.8% of revenues).

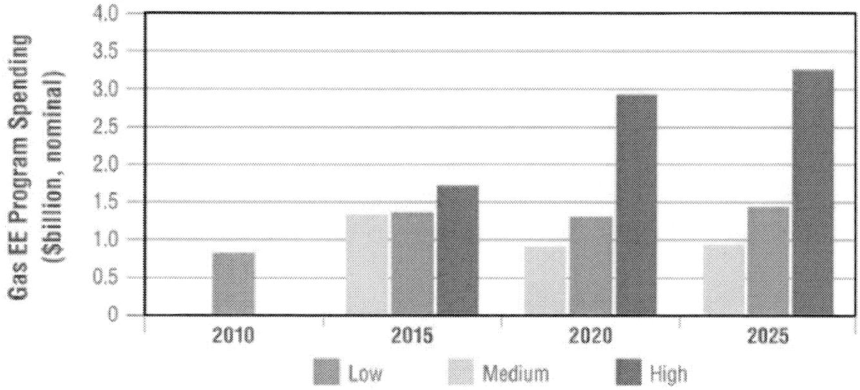

Figure 7. Projected Gas Energy Efficiency Program Spending.

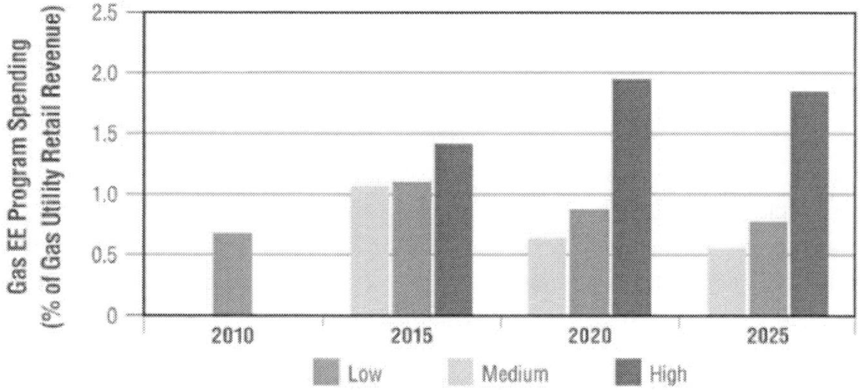

Figure 8. Projected Gas Energy Efficiency Program Spending as a Percentage of Gas Distribution Utility Revenues.

5. Discussion of Key Issues and Uncertainties

The preceding set of projections suggest a wide range of potential trajectories for utility customer-funded energy efficiency program spending and savings in the United States – even without considering the possibility of fundamentally new policy developments. In this section, we identify some of the significant issues and uncertainties that may influence the spending course and impact of these programs over the next 10 to 15 years and which we attempted to account for – either directly or indirectly – within the projections. These interrelated issues and uncertainties include both external factors, such as the broader policy and market context within which utility customer-funded programs operate, and internal factors related to the implementation and regulatory oversight of these programs.

5.1. Broader Market and Policy Context

Utility customer-funded energy efficiency programs and their enabling policies function within a broader context, influenced by a variety of market forces and conditions, as well as by interactions with other policies. We briefly highlight four key elements of this broader market and policy context that may be particularly critical to the future trajectory of customer-funded efficiency programs: the state of the economy, natural gas prices, federal minimum efficiency standards, and environmental regulations affecting the electric power sector.[22]

The Economy

The timing and extent of the economic recovery may complicate and restrain efforts to scale-up energy efficiency spending and savings over the near to medium term, for several reasons. First, utility customer-funded energy efficiency programs typically requires customers to pay a portion of the capital outlay for energy efficiency measures; as households and businesses struggle to manage their day-to-day expenses, and as declining home values reduce the equity available for financing efficiency improvements, many customers may be reluctant to make new investments, even those with short payback periods. As a result, program participation may be suppressed, or program costs may rise if program administrators are required to increase financial incentives or expend greater sums on marketing efforts. Second, a stagnant economy is likely to reduce the rate of stock turnover and new housing starts, thereby reducing the amount of energy savings that could be captured through utility customer-funded programs targeting these market opportunities. Third, a slow economy may indirectly constrain energy efficiency program efforts in at least three ways: heightened sensitivity to potential near-term rate impacts associated with efficiency program spending,[23] increased risk that policymakers will re-direct dedicated funding for energy efficiency to shore-up state budgets[24] or other non-efficiency purposes, and slowed load growth, thereby reducing the avoided capacity costs and cost effectiveness of energy efficiency programs.

Moderate Natural Gas Prices

As of April 2012, natural gas was trading at wellhead prices of less than $2 per million British thermal units (MMBtu), the lowest level in 10 years and nearing a record low. Although natural gas prices are projected to rise over the next 20 years (EIA 2012), they are nevertheless expected to remain lower, in real terms, than the prices that characterized most of the past decade, when most state energy savings targets were set.[25] For electric and gas energy efficiency programs, lower gas prices translate into reduced program benefits, which in turn constrains total efficiency spending and flexibility in program design as benefit-cost ratios decrease. More aggressive efficiency portfolios and comprehensive, multi-measure programs may be especially at risk, because costlier measures will result in longer payback periods for customers and will not be as cost effective from a total resource cost perspective. The effects of moderate gas prices will be especially pronounced for natural gas efficiency programs because lower gas commodity costs means lower avoided energy costs to gas utilities, which affects program cost effectiveness. Lower gas prices also mean that customers will have incentive to increase consumption or convert to gas heating from other fuels and will have less direct financial incentive to invest in energy efficiency.

State and Federal End-Use Codes and Standards

In recent years, state adoptions of building energy codes have increased, and federal minimum efficiency standards for appliances and end-use equipment have been tightened. These policies affect utility customer-funded programs by essentially raising the baseline against which savings are measured, thereby influencing both the size of the remaining potential that can be harvested through those programs and the mix of technologies targeted. Two specific federal efficiency standards that are planned to go into effect over the near-term – for lighting in 2012 to 2014, then again in 2020, and for non-weatherized natural gas furnaces in 2013 – may have potentially significant impacts on customer-funded efficiency programs. The impact of the federal lighting standards is somewhat less certain, because program administrators have other lighting technologies that are likely to remain cost effective after the standards come into effect. Gas program administrators, however, may have fewer options. Starting in 2013, the new furnace standards would raise the minimum heat-to-fuel efficiency of furnaces from 78% to 90% AFUE[26] in northern states (generally the states with the nation's most substantial spending and savings targets). Programs can continue to provide incentives for higher efficiency gas furnaces, but with a technological efficiency limit of about 98% AFUE, the incremental savings will be lower, and residential gas furnace programs are therefore less likely to continue as the mainstay of gas efficiency program portfolios.

Emissions Regulations

Proposed or final air emissions regulations that are being considered or adopted by state and federal environmental agencies[27]- in combination with low-priced, abundant gas – have become important drivers for utility customer-funded energy efficiency programs, as part of utilities' multi-faceted strategies for managing the retirement of older coal-fired generators.[28] For example, many utility resource plans have discussed the potential role of demand-side resources as part of a strategy for complying with emissions requirements (e.g., Tennessee Valley Authority), as a prerequisite for utility customer funding of low carbon replacement generation (American Electric Power in West Virginia, Florida Power & Light in Florida), or as a means of deferring retirement and replacement decisions (Duke Energy Carolinas). The ultimate import of these regulations for future energy efficiency program budgets, however, depends on several factors. These factors include: the timing and stringency of the final rules; the price of natural gas (as gas-fired generation is expected to offset the majority of the retired coal-fired generation); the capital cost profile of clean energy generation alternatives (e.g., renewable energy, nuclear power, coal with carbon capture and sequestration); the regulatory and business models in place that govern the balance and relative attractiveness of supply- and demand-side investments; and the degree to which utilities and utility regulators integrate state and tribal Clean Air Act implementation plans with utility resource plans.

5.2. Energy Efficiency Program Implementation and Regulatory Oversight

There are also a variety of other critical issues and uncertainties specific to the regulatory and administrative institutions within which utility customer-funded efficiency programs operate and that may strongly influence the spending and savings trajectories of those programs. Here, we highlight several: general aversion to rate impacts, challenges associated with developing innovative program designs to reach deeper and broader savings, and the limited ability in some states to extend gas efficiency programs to transportation gas customers.

Aversion to Rate Impacts

In most states, utilities typically expense program costs for energy efficiency as they are incurred. As a result, energy efficiency program cost recovery is relatively front-loaded compared to cost recovery for most utility supply-side resource alternatives. As a result, the rate impacts from energy efficiency tend to occur sooner (even if the rate impacts are less over the long-term, and even if average utility bills are reduced compared to supply-side alternatives). The short-term rate impacts associated with attaining very aggressive levels of savings (or even relatively modest levels of savings in states that are higher than has historically occurred) could pose a political challenge for state regulators, particularly in states that have seen significant rate hikes in recent years or whose rates are well-above national averages. Across all states, these challenges are further heightened during periods of economic hardship. Concerns about rate impacts from energy efficiency programs have been institutionalized in a number of states, either through explicit caps on spending or rate impacts, or by the application of the ratepayer impact measure (RIM) test.[29] Meeting aggressive EERS targets in some states will likely require exceeding these caps or otherwise justifying rate increases, which may be feasible only in a robust, growing economy.

Developing Innovative Program Designs to Reach Deeper and Broader Savings

A number of states have established aggressive energy efficiency savings goals for future years that are well beyond current experience and practice in most leading states (e.g., annual incremental electric savings on the order of 1.5% to 2% or more of retail sales). The challenge for these program administrators will be to design and implement programs that can achieve both deeper savings, on average, at customer facilities and have a broader reach in terms of market penetration over a sustained period of time. Service providers will have to achieve savings levels of 25-40% of existing usage at customer facilities compared to current practice in utility customer-funded programs, which is typically in the 5-20% range. Achieving higher market penetration rates will require programs to target and reach traditionally under-served markets (e.g., small commercial, multi-family, rental housing, moderate income households, non-owner occupied commercial facilities) in far greater numbers than current practice (MEEAC 2009). We are also likely to see increased attention to integrated delivery of electric and gas efficiency programs as well as coordinated delivery of energy efficiency, on-site renewables and combined heat and power, in order to reduce transaction costs and provide customers with tailored, customized service offerings.

Extending Programs to Transportation Gas Customers

In many states, energy savings in the large commercial and industrial markets are, in effect, beyond the reach of program administrators. This is especially true for gas efficiency programs, as large commercial and industrial customers often purchase natural gas on the competitive market through alternative retailers, and may not pay into or be able to participate in gas utility customer-funded energy efficiency programs.[30] This "transportation gas" accounts for 46% of total U.S. gas sales and 79% of all commercial and industrial sales.

The ability for many states to significantly increase gas efficiency program savings and spending may therefore hinge, to a large degree, on whether mechanisms can be developed (e.g., non-bypassable charges for program funding) to bring these customers and savings opportunities into the program fold.

CONCLUSION

Energy efficiency programs funded by utility customers are poised for dramatic growth over the course of the next 10 to 15 years, especially in the Midwest and South – with a contingent of populous Midwest states ramping up to meet statutory EERS targets, and in the South, the expectation that a collection of relatively modest EERS policies and nascent IRP/DSM planning processes in states with a large base of energy consumption will push spending upward from currently low levels. As a result, program spending is expected to become more evenly distributed nationwide by 2025.

Program spending is projected to roughly double to $9.5 billion in 2025 and could reach $15.6 billion under aggressive assumptions about the policy support, implementation and effectiveness of current policies. Program administrators in many states are projected to achieve annual electricity savings of between 1.5% and 2%, surpassing the achievements of most leading states today.

Given forecasts for a slow economy recovery and modest load growth, the projected growth in electricity program spending and savings under our medium case scenario would offset most aggregate annual U.S. electric load growth through 2025, based on the load forecast in EIA's most recent reference case (and given assumptions about the extent to which EIA's forecast captures the impact of future efficiency programs).

However, program administrators and state regulators face emerging challenges and uncertainties. The combined effects of economic torpor, moderate gas prices, and tightening energy codes and minimum efficiency standards pose challenges for continued growth in electric and, especially, gas efficiency programs.

The degree to which leading states and a new vanguard of fast-rising states can overcome these challenges and offset reduced efforts elsewhere is likely to govern the longer term path for national-level spending and savings on efficiency programs.

REFERENCES

Barbose, G., C. Goldman, and J. Schlegel. 2009. The Shifting Landscape of Ratepayer-Funded Energy Efficiency in the U.S. Berkeley, CA: Lawrence Berkeley National Laboratory. LBNL2258E. http://eetd.lbl.gov/ea/emp/reports/lbnl-2258e.pdf

Bipartisan Policy Center. 2011. Environmental Regulation and Electric System Reliability. http://www.bipartisanpolicy.org/library/report/environmental-regulation-and-electric-systemreliability

Boonekamp, Piet G. M. 2011. How much will the energy service directive contribute to the 20% EU energy savings goal? *Energy Efficiency 4:85-301, DOI: 10.1007/s12053-010-9088-0*

Brattle Group. 2010. Potential Coal Plant Retirements Under Emerging Environmental Regulations. http://www.brattle.com/_documents/upload library/upload898.pdf

Congressional Research Service. 2011. EPA's Regulation of Coal-Fired Power: Is a `Train Wreck' Coming? CRS-R41914

Consortium for Energy Efficiency (CEE). 2012. State of the Efficiency Program Industry: Budgets, Expenditures and Impacts 2011. http://www.cee1.org/files/2011%20CEE%20Annual%20Industry%20Report.pdf

Consortium for Energy Efficiency (CEE). 2008. Energy Efficiency Programs: A $3.7Billion U.S. and Canadian Industry. 2007 Report. (http://www.cee1.org/eepe/2007/2007EEPReport.pdf)

Cooper, A. and L. Wood. 2012. Summary of Ratepayer-Funded Electric Efficiency Impacts, Budgets, and Expenditures (2010-2011). Washington, D.C.: Institute for Electric Efficiency. http://www.edisonfoundation.net/iee/Documents/IEE CEE2011 FINAL update.pdf

Dixon, R.K., E. McGowan, G. Onysko,, R.M. Scheer.US energy conservation and efficiency policies: Challenges and opportunities, Energy Policy, Volume 38, Issue 11, November 2010, Pages 6398-6408, ISSN 0301-4215, 10.1016/j.enpol.2010.01.038. (http://www.sciencedirect.com/science/article/pii/S0301421510000637)

Eldridge, M., M. Neubauer, D. York, S. Vaidyanathan, A. Chittum, and S. Nadel. 2008. The 2008 State Energy Efficiency Scorecard. Washington, D.C.: American Council for an Energy Efficient Economy. http://www.aceee.org/pubs/e086.htm.

Energy Information Administration (EIA). 2012. Annual Energy Outlook, 2012. (Accessed June 2012)

EPRI. 2009. *Assessment of Achievable Potential from Energy Efficiency and Demand Response Programs in the U.S.: (2010–2030)*. EPRI, Palo Alto, CA: 2009. 1016987.

Foster, B., A. Chittum, S. Hayes, M. Neubauer, S. Nowak, S. Vaidyanathan, K. Farley, K. Schultz, and T. Sullivan. 2012. *The 2012 State Energy Efficiency Scorecard.* ACEEE Research Report EC12. http://aceee.org/research-report/e12c

Geller, H., P. Harrington, M.D. Levine, A.H. Rosenfeld and S. Tanishima. 2006. Policies for Increasing Energy Efficiency: Thirty Years of Experience in OECD Countries. *Energy Policy* 34(5): 556–573.

Gillingham, K., R. Newell, and K. Palmer. June 2004, revised September 2004. Retrospective Examination of Demand-Side Energy Efficiency

Policies. RFF DP 04-19 REV. http://www.rff.org/RFF/Documents/RFF-DP-04-19REV.pdf.

Goldman, C., E. Stuart, I. Hoffman, M. Fuller and M. Billingsley. 2011. Interactions Between Energy Efficiency Programs Funded Under the Recovery Act and Utility Customer-Funded Energy Efficiency Programs. Lawrence Berkeley National Laboratory, LBNL-4322E. http://eetd.lbl.gov/EA/EMP/reports/lbnl-4322e.pdf.

Hopper, Nicole, Galen Barbose, Charles Goldman, and Jeff Schlegel. 2008. "Energy Efficiency as a Preferred Resource: Evidence from Utility Resource Plans in the Western United States and Canada." *Energy Efficiency*, 14(2):126-34, October 2008. http://www.springerlink.com /content/6k46076226625763.

MEEAC (Massachusetts Energy Efficiency Advisory Council). 2009. The Context for Energy Efficiency Savings: Electric Savings Update, Electric Net Benefits and Gas Savings. Presentation by Council Consultants at MEEAC meeting. March 24, 2009. http://www.ma- eeac.org.

Nadel, S.. 1992. Utility Demand-Side Management Experience and Potential - A Critical Review. *Annual Review of Energy and the Environment* 17:507-535. http://www.aceee.org/sites/default/files/publications /research reports/U921.pdf

Navigant Consulting Inc. 2012. Analysis To Update Energy Efficiency Potential, Goals, And Targets For 2013 And Beyond. Prepared for the California Public Utilities Commission. http://www.cpuc.ca.gov/NR /rdonlyres/5A1B455F-CC46-4B8D-A1AF-34FAAF93095A/0/2011IOUServiceTerritoryEEPotentialStudyFinalReport.pdf

Nowak, S., M. Kushler, M. Sciortino, D. York, and P. Witte. 2011. Energy Efficiency resource Standards: State and Utility Strategies for Higher Energy Svaings. ACEEE Research Report U113. http://www.aceee.org /research-report/u113

Sciortino, M., M. Neubauer, S. Vaidyanathan, A. Chittum, S. Hayes, S. Nowak, and M. Molina. 2011. The 2011 State Energy Efficiency Scorecard. ACEEE Research Report E115. http://www.aceee.org /research-report/e115

York, D., P. Witte, K. Friedrich, M. Kushler. 2012. A National Review of Natural Gas Energy Efficiency Programs. ACEEE Report Number U121. (http://www.aceee.org/sites/default/files/publications/researchreports/u121 .pdf)

Technical Appendix A: Methodology and Assumptions Used to Develop Energy Efficiency Spending and Savings Projections

This technical appendix describes the methods and assumptions used to develop projections of energy efficiency program spending by U.S. electric and natural gas customers and savings (electric only) through 2025.

A1. Electric Energy Efficiency Spending & Savings Projections

Low, medium, and high projections of future electric energy efficiency program savings and spending were developed on a state-by-state basis. Although many of the specific assumptions and the approach to defining scenarios varied by state, the basic methodology used in all states consisted of several common components, including:

- Developing projections of retail electricity sales and revenues from retail electricity sales;
- Defining low, medium, and high scenarios of future utility customer-funded energy; efficiency program savings and spending for the electricity sector; and
- Estimating the amount of spending required to achieve different levels of savings.

Each of these elements is described further below.

A1.1. Retail Sales and Revenue Projections

Projections of annual retail electricity sales and revenues were used as an input to develop energy efficiency program savings and spending projections; for example, in those states that establish EERS targets as a percent of retail sales, savings projections were calculated based on projected retail sales. Sales and revenue projections were also used to develop metrics that allow for comparison of spending and savings levels across states of differing sizes (e.g., savings as a percent of retail sales and spending as a percent of revenues).

Baseline Retail Sales and Revenue Projections

An initial set of baseline retail sales and retail price projections for each state was developed by applying annual growth rate projections from the Energy Information Administration (EIA)'s 2012 Annual Energy Outlook (AEO2012) reference case forecast to actual 2010 retail sales and price data for each state, as reported on EIA's Form-860. The electricity retail sales and price projections in AEO2012 are specified at the Electricity Market Module (EMM) level, the regions used in EIA's National Energy Modeling System (NEMS); the state-level retail electricity sales and average retail electricity price projections were developed by applying the EMM-level growth rates to historical retail sales and revenues for each state in the respective region. Table A-1 summarizes the annual average growth rates (2011 to 2025) of retail electricity sales and electricity prices in each EMM region from the AEO2012 reference case forecast. Revenue projections were calculated by multiplying projected retail electricity prices by projected retail electricity sales.

Table A - 1. AEO Projected Growth Rates in Retail Electricity Sales and Prices

Electricity Market Model (EMM) Region	States	AEO2012 Average Annual Growth Rate (2011-2025)	
		Retail Electricity Sales	Retail Electricity Prices Price (nominal)
Electric Reliability Council of Texas	TX	0.7%	2.0%
Florida Reliability Coordinating Council	FL	0.9%	1.6%
Midwest Reliability Council / East	WI	0.1%	1.5%
Midwest Reliability Council / West	IA, MN, ND, NE, SD	0.2%	1.6%
Northeast Power Coordinating Council / Northeast	CT, MA, ME, NH, RI, VT	0.2%	1.2%
Northeast Power Coordinating Council / NYC-Westchester	NY	0.3%	1.1%
Northeast Power Coordinating Council / Long Island	NY	0.1%	1.4%
Northeast Power Coordinating Council / Upstate New York	NY	0.0%	1.4%
Reliability First Corporation / East	DC, DE, MD, NJ, PA	0.3%	1.4%
Reliability First Corporation / Michigan	MI	0.1%	1.5%

Table A - 1. (Continued)

Electricity Market Model (EMM) Region	States	AEO2012 Average Annual Growth Rate (2011-2025)	
		Retail Electricity Sales	Retail Electricity Prices Price (nominal)
Reliability First Corporation / West	IL, IN, OH, WV	0.2%	2.4%
SERC Reliability Corporation / Delta	AR, LA, MS	0.7%	2.2%
SERC Reliability Corporation / Gateway	MO	0.2%	2.2%
SERC Reliability Corporation / Southeastern	AL, GA	0.8%	1.4%
SERC Reliability Corporation / Central	KY, TN	0.8%	1.0%
SERC Reliability Corporation / Virginia-Carolina	NC, SC, VA	0.9%	1.4%
Southwest Power Pool / North	KS	0.3%	2.1%
Southwest Power Pool / South	OK	0.7%	2.3%
Western Electricity Coordinating Council / Southwest	AZ, NM, NV	1.3%	2.0%
Western Electricity Coordinating Council / California	CA	0.8%	1.9%
Western Electricity Coordinating Council / Northwest Power Pool Area	ID, MT, OR, WA, WY	0.9%	0.4%
Western Electricity Coordinating Council / Rockies	CO, UT	1.3%	1.8%

State-Specific Adjustments for Each Scenario

Future retail sales and revenues in each state will depend, in part, on the amount of savings achieved from future customer-funded energy efficiency programs. In order to maintain internal consistency, we adjusted the retail sales and revenue projections for each scenario in each state, to reflect the energy efficiency savings assumed for the given scenario. The adjustments consisted of decreasing (or increasing) the baseline sales and revenue in each year, to account for the cumulative difference between the savings assumed for the scenario and the savings assumed to be implicit in the AEO2012 forecast. To provide an example: if we project, under one scenario, that future annual incremental savings in a given state will be equal to 0.3% of retail sales in each year, and the energy efficiency savings assumed to be implicit in the baseline retail sales forecast are 0.1% of retail sales, then we would reduce the forecast in each year to account for the cumulative effect of the additional 0.2% of retail sales saved each year (i.e., reduce the retail sales projection by 0.2% in year one, by 0.4% in year two, and by 0.6% in year three and so on).

The foregoing adjustment requires an estimate of the savings embedded in the AEO-derived baseline retail sales forecast for each state. Although NEMS does not explicitly account for the impacts of future utility customer-supported efficiency programs, the model operates under the implicit assumption that historical trends in utility customer-funded efficiency programs will continue over the forecast period.[31] We therefore assumed that the baseline retail sales projections, derived from the AEO2012 forecasted growth rates, implicitly account for a continuation of customer-funded energy efficiency program savings equal to the average level achieved over the 2000 to 2010 period. Comprehensive and reliable state-level data on efficiency program savings are available only from 2006 onward, although sufficiently reliable historical spending data for the U.S. as a whole are available going back to 2000. Thus, the average annual incremental savings in each EMM region over the 2000-2010 period were calculated via a two-step process. First, historical state-level savings from 2006 to 2009 were aggregated up to each EMM region to calculate the average annual incremental savings in each EMM region over that period. Second, the 2000-2010 average EMM savings were calculated by multiplying the 2006-2009 average EMM savings (from the previous step) by the ratio of average U.S. spending over 2000 to 2010 to average U.S. spending over 2006 to 2009. The result of this two-step calculation yielded the annual incremental energy efficiency program savings assumed to be embedded in the AEO-derived baseline retail sales forecast for each EMM region (see Table A-2).

Table A - 2. Estimate Customer-Funded Energy Efficiency Program Savings Embedded in AEO Forecasts by EMM Region

Electricity Market Model (EMM) Region	Annual Incremental Savings (% of Retail Sales)
Electric Reliability Council of Texas	0.1%
Florida Reliability Coordinating Council	0.1%
Midwest Reliability Council / East	0.4%
Midwest Reliability Council / West	0.3%
Northeast Power Coordinating Council / Northeast	0.6%
Northeast Power Coordinating Council / NYC-Westchester	0.3%
Northeast Power Coordinating Council / Long Island	0.3%

Table A - 2. (Continued)

Electricity Market Model (EMM) Region	Annual Incremental Savings (% of Retail Sales)
Northeast Power Coordinating Council / Upstate New York	0.3%
Reliability First Corporation / East	0.1%
Reliability First Corporation / Michigan	0.1%
Reliability First Corporation / West	0.0%
SERC Reliability Corporation / Delta	0.0%
SERC Reliability Corporation / Gateway	0.0%
SERC Reliability Corporation / Southeastern	0.0%
SERC Reliability Corporation / Central	0.0%
SERC Reliability Corporation / Virginia-Carolina	0.1%
Southwest Power Pool / North	0.0%
Southwest Power Pool / South	0.0%
Western Electricity Coordinating Council / Southwest	0.3%
Western Electricity Coordinating Council / California	0.6%
Western Electricity Coordinating Council / Northwest Power Pool Area	0.3%
Western Electricity Coordinating Council / Rockies	0.3%

A1.2. Scenario Definitions

Scenarios were defined on a state-by-state basis, driven primarily by each state's unique policies, regulatory dynamics, program histories, available reports and approved DSM or IRP plans. In order to simplify the scenario development and characterization process, we grouped states into Census Regions (see Figure A-1). Tables A-3 through A-6 summarize the assumptions used for each state by region. For scenarios defined in terms of assumed savings, spending levels are derived from savings, and vice-versa. For a group of seven "uncommitted" states that currently have little efficiency program activity and no established policy framework, we instead employed a standardized approach by which spending as a percentage of revenues was stipulated to increase above current levels by a specified amount under each scenario, as described in Table A-7.[32] Unless otherwise indicated, spending and savings projections for municipal utilities and cooperatives within the states covered in Tables A-3 through A-6 are also developed using the same assumptions as for uncommitted states.

Figure A - 1. U.S. Census Regions Used for Scenario Development.

Table A - 3. Scenario Descriptions for States in the Northeast

State	Case	Scenario Assumptions
CT	Low	SBC funding remains at current approved levels, but RGGI and FCM funding are no longer used for utility customer-funded energy efficiency.
	Medium	Spending percentage remains at current levels indefinitely; RGGI and FCM funding remains at current levels over entirety of forecast period.
	High	IOU savings equal to all achievable energy efficiency potential, based on NEEP potential analysis, but subject to an annual spending cap equal to 6.5% of revenue
MA	Low	IOU spending/savings from 2010 to 2012 based on most recent plans. For subsequent years, utility customer funding declines to the legislatively mandated SBC plus current RGGI/FCM levels (e.g., consistent with a future in which very aggressive codes and standards exhaust much of the cost-effective energy efficiency potential). POU spending/savings based on generic assumptions in all scenarios.

Table A - 3. (Continued)

State	Case	Scenario Assumptions
	Medium	IOU savings levels remain flat at 2012 level, equal to 2.3% percent of retail sales.
	High	IOUs meet state policy of acquiring all cost-effective energy efficiency, with savings estimated based on the energy efficiency potential study conducted by NEEP, subject to an assumed annual spending cap equal to 10% of revenue from retail sales.
ME	Low	Spending remains constant at 2011 level of 1.6% of revenues from retail sales
	Medium	Spending from 2011 to 2013 based on Efficiency Maine's projected budget, and remains at 2013 level, equal to 2.9% of revenues from retail sales, through 2025.
	High	Savings rise to achieve all cost-effective energy efficiency, based on NEEP analysis, but subject to a spending cap equal to 3.6% of revenues.
NH	Low	Spending levels through 2020 are based on a continuation of planned 2012 SBC & FCM funding levels for electric energy efficiency, equal to 1.3% of revenues from statewide electricity retail sales.
	Medium	Savings ramp up from 0.6% of retail sales in 2012 (the projected savings under current approved funding levels) to 1.2% of retail sales in 2020, and remain constant at that level thereafter.
	High	Same as the Medium Case, except savings ramp up to 2.0% of retail sales in 2020.
NJ	Low	Spending from 2013 to 2016 is based on RFP (strawman) issued by BPU for a new program administrator, de-rated to 70% to reflect historical levels of under-spending. After 2016, spending on incentive-based programs remains flat at the 2016 level, and spending on financing programs remains at the average of the projected level over the 2013-2016 period.
	Medium	Spending from 2013 to 2016 is based on RFP (strawman) issued by BPU for a new program administrator, with no de-rating as is applied in the Low Case.After 2016, spending on both incentive-based programs and financing programs remains flat at the projected 2016 level.
	High	Same as Medium Case through 2016, and after 2016, spending on financing programs is also the same as the Medium Case. However, spending on incentive-based programs after 2016 rises such that total savings from both incentive-based programs and financing programs reaches 1.5% of retail sales in 2025.

State	Case	Scenario Assumptions
NY	Low	Funding remains at 2010 levels at 2.5% revenues.
	Medium	Savings from 2009 to 2015 are based on the portion of the state's overall EEPS target allocated by the NYPSC to utility customer-funded programs. Spending from 2016 to 2020 remains constant at the average level during 2009 to 2015, equal to 2.9% of revenues.
	High	Savings from 2009 to 2015 are based on the portion of the state's overall EEPS target allocated by the NYPSC to utility customer-funded programs. Spending from 2016 to 2020 remains constant at 3.9% of revenues.
PA	Low	Spending through 2012 is based on approved budget, and decreases to 75% of 2012 spending in 2013, equal to 1.5% of revenue
	Medium	Spending rises to 2% of 2008 revenue, equal to $268 million (current cap) in 2013 and continues indefinitely
	High	Same as Low/Medium case through 2013, but then rate cap is removed and IOU savings increase to 1.0% of retail sales in 2025
RI	Low	Spending through 2025 remains flat at the 2011 level proposed by National Grid in the approved settlement for the 2011 Energy Efficiency Plan, estimated to be 4.3% of revenues.
	Medium	Savings ramp up from planned 1.3% of retail sales in 2011 to 1.9% in 2025, as achieved by MA IOUs under Medium Case.
	High	IOUs acquire all cost-effective energy efficiency, with savings estimated based on the energy efficiency potential study conducted by NEEP, subject to an assumed annual spending cap equal to 10% of revenue from retail sales.
VT	Low	Spending through 2014 is based on Efficiency Vermont's approved 2012-2014 budgets plus Burlington Electric's 2010 spending; spending from 2015 onward remains constant at the average of 2012-2014 levels, equal to 6.7% of statewide revenues.
	Medium	Same as Low Case
	High	Acquire all cost-effective EE, with savings estimated based on the EE potential study conducted by NEEP, subject to an assumed annual spending cap equal to 10% of revenue from retail sales.

Table A - 4. Scenario Descriptions for States in the South

State	Case	Scenario Assumptions
AL	Low	TVA savings based on 2010 IRP "Baseline Portfolio," rising to 0.3% of retail sales. Non-TVA load based on "uncommitted state" low scenario.
	Medium	TVA savings based on 2010 IRP selected portfolio, rising to 0.7% of retail sales. Non-TVA load based on "uncommitted state" medium scenario.
	High	TVA savings based on 2010 IRP "High EEDR" portfolio, rising to 1.0% of retail sales. Non-TVA load based on "uncommitted state" state high scenario.
AR	Low	Annual savings from IOU utility customer-funded energy efficiency through 2013 are based on most recently approved DSM plans, and savings remain flat at 2013 levels (0.8% of retail sales) thereafter.
	Medium	Same as Low Case, except IOU savings rise to 1.0% annual savings in 2017 and remain flat at that level thereafter.
	High	Same as Low Case, except IOU savings rise to 1.25% annual savings in 2021 and remain flat at that level thereafter.
DC	Low	The DC City Council terminated all but about $2 million in funding for 2011 programs. Energy efficiency spending drops that year then ramps gradually to 1% of revenues in 2020 and remains flat to 2025.
	Medium	Savings ramp up to 1% of retail sales in 2013 (a year later than mandated in the District contract with VEIC) and remains flat to 2025.
	High	Savings ramp up to 1% in 2013 as in the Medium Case, then to 1.2% of retail sales in 2020, then flat to 2025
DE	Low	The Sustainable Energy Utility ceases to exist in 2012. No utility customer funding for energy efficiency until utilities start their own programs in 2020 (next IRP, assumed funding is equal to RGGI funding).
	Medium	SEU continues to exist funded by the proceeds of the RGGI auctions (assumed constant at 2011 levels).
	High	SEU contines to exist with RGGI funding, utilities start their own programs in addition to the SEU. Savings achieve 1% of retail sales in 2020.
FL	Low	IOUs and large munis achieve savings in their site plans, which fall to 0.14% of retail sales in 2021, and remain flat at that level to 2025. Remaining utilities based on "uncommitted state" low case assumptions.

State	Case	Scenario Assumptions
	Medium	Savings for IOUs and large munis in each year are equal to the mid-point of the low and high cases. Remaining utilities based on "uncommitted state" medium case assumptions through 2019, reaching savings of 0.3% of retail sales and remaining at that level thereafter.
	High	Savings for IOUs and large munis ramp from current levels of 0.2% to 0.5% in 2019, per the FPUC's adopted 2010-2019 goals, and then ramp up further to 0.75% in 2025. Remaining utilities based on "uncommitted state" high case assumptions.
GA	Low	GA Power savings through 2013 are based on approved 2011-2013 DSM plan, and remain flat at 2013 level thereafter (0.3% of retail sales).
	Medium	GA Power savings based on 2010 IRP targets, which rise to 0.4% of retail sales by 2020, and are assumed to remain flat thereafter.
	High	GA Power savings assumed to ramp up to 1.0% of retail sales by 2025.
KY	Low	IOUs realize early-year savings from 2011 DSM plan and 2011 IRP, reaching 0.4% of retail sales in 2013 and flat thereafter. For TVA load, see Alabama (AL).
	Medium	IOUs realize savings from DSM portfolio and 2010 IRP, reaching 0.6% of retail sales in 2014 and constant thereafter. For TVA load, see Alabama (AL).
	High	IOUs realize highest level of savings from 2011 DSM plans and 2011 IRP, reaching 0.8% of retail sales in 2014 and constant thereafter. For TVA load, see Alabama (AL).
MD	Low	IOU and SMECO spending through 2015 based on their proposed 2012 energy efficiency plans, but with a 1-year lag. For 2015 to 2020, spending remains flat at 2015 level, equal to 2.3% of revenues.
	Medium	IOU and SMECO spending through 2014 based on their proposed 2012 energy efficiency plans. For 2015 to 2020, spending remains flat at the 2015 level, equal to 2.3% of revenues.
	High	IOU and SMECO achieve the EmPower Maryland energy efficiency savings target in 2015. For 2015 to 2020, savings levels stay at 1.6% of retail sales.
MS	Low	Same as Alabama (AL)
	Medium	Same as Alabama (AL)
	High	Same as Alabama (AL)
NC	Low	IOUs exhaust, but do not exceed, the energy efficiency allowance under the state RPS; POUs first exhaust their allowable use of large hydro, and then meet 50% of remaining RPS needs (after all set-asides are met) with energy efficiency.

Table A - 4. (Continued)

State	Case	Scenario Assumptions
	Medium	Use the Baseline EE Scenario adopted by the NC. Energy Policy Commission (based upon the La Capra Associates report).
	High	All utilities ramp up from annual savings levels of 0.3% of retail sales in 2010 (as projected for that year in the Low/Medium Cases) to 1.3% in 2020, then flat through 2025. This is largely consistent with the NC Energy Policy Commission's high energy efficiency scenario, which is based upon $0.10/kWh avoided costs.
SC	Low	Majority of savings and spending are SC share of NC REPS-driven savings and spending by Duke and Progress. Extend IOU savings at 2013 levels (0.5% of retail sales) through 2025.
	Medium	Assume Duke, Progress and SCEG realize the base case savings projections from their IRPs, rising to 0.7% of retail sales by 2020.
	High	Assume Duke meets its IRP High Case and other large IOUs take their IRP base case savings projection, with total IOU savings rising to 0.8% of retail sales in 2020 and ramping up further to 1.1% of retail sales in 2025.
TN	Low	Same as Alabama (AL)
	Medium	Same as Alabama (AL)
	High	Same as Alabama (AL)
TX	Low	IOU savings meet current EERS goal (20% of incremental peak demand, ramping to 25% in 2012 and 30% from 2013 on); POU savings/spending are based on generic low case assumptions.
	Medium	IOU savings assumed to continue at present levels of over-performance relative to EERS goal. POU savings/spending are based on the generic high case assumptions, with savings ramping up to 0.5% of retail sales by 2025.
	High	Statewide savings are based on both IOUs and POUs meeting the current EERS savings targets through 2014 (as in the Mid-Case), with savings from 2015 onward equal to 50% of demand growth in each year (as previously proposed).
VA	Low	Spending as a percent of revenues by Dominion Virginia Power continues at the current rate (0.2% of revenues).
	Medium	IOUs reach 0.4% annual savings in 2018 and remain flat thereafter, based roughly on the spending trajectory from Dominion Virginia's most recent IRP.
	High	All IOUs meet 10% by 2022 voluntary EERS (relative to 2006 load).

Table A - 5. Scenario Descriptions for States in the Midwest

State	Case	Scenario Assumptions
IA	Low	Statewide spending (IOUs & POUs) continues indefinitely at 2010 approved spending levels, 2.9% of revenue, with corresponding savings roughly constant at 1.0% of retail sales.
	Medium	For 2009 to 2013, IOU spending projections are based on the approved budgets in their 2009-2013 energy efficiency plans, and POU energy efficiency spending is equal to the projected annual spending for 2010 to 2013. For 2014 to 2020, continue IOU & POU spending at projected 2013 level 3.4% of revenues, with corresponding savings roughly constant at 1.2% of retail sales.
	High	All utilities (IOUs & POUs) achieve savings equal to 1.5% of retail sales by 2011 (the target that IOUs were required to evaluate in their most recent energy efficiency plans) and continue at that level until 2020 when they reach 2% savings.
IL	Low	IOU savings/spending through 2013 are based on approved energy efficiency plans. From 2014 to 2020, savings equal EEPS targets until statutory cost caps are reached, which limits savings to approximately 0.8% of retail sales.
	Medium	Same as Low Case, except IOU savings are assumed to rise to 1.3% of retail sales in 2016
	High	IOU savings/spending through 2013 are based on approved energy efficiency plans. Statutory EEPS spending cap is assumed to be lifted or increased, and EEPS targets are assumed to be fully achieved.
IN	Low	IURC-jurisdictional utilities (IOUs and some coops and munis) achieve savings levels equal to 75% of EERS levels.
	Medium	IURC-jurisdictional utilities fully meet EERS savings targets.
	High	Same as Medium Case.
MI	Low	Statewide savings based on full compliance with statutory EEPS targets, which reach 1.0% of retail sales in 2012.
	Medium	Same as Low Case.
	High	Assumes that statewide savings ramp up from 1.0% of retail sales in 2012 to 1.4% in 2020.
MN	Low	All utilities meet the minimum 1.0% EEPS target for conservation improvement programs.
	Medium	Savings ramp up from 1.0% of retail sales in 2010 to 1.2% in 2020.
	High	Same as Medium Case, except savings ramp up to 2.0% in 2020.

Table A - 5. (Continued)

State	Case	Scenario Assumptions
MO	Low	IOU annual savings are based on most recently approved IRPs and DSM plans through 2015, and savings thereafter remain flat at that level.
	Medium	Same as the Low Case through 2015; from 2016 onward, IOU savings as a percentage of retail sales are the mid-point between the Low and High Cases.
	High	IOU savings track approved/submitted plans through 2014, then meet voluntary goals in MO PSC's MEEIA regulations, rising to 1.9% of retail sales in 2020 and remaining at that level thereafter.
OK	Low	IOU savings and spending through 2015 based on most recent filed plans, and savings thereafter remain flat at the 2015 level.
	Medium	Same as Low Case through 2015, after which IOU savings ramp up to 0.75% of retail sales in 2020 (as a proxy for the soft rate cap on residential DSM charges) and remain flat at that level thereafter.
	High	Same as Medium Case through 2020, but IOU savings continue to ramp up to 1.0% of retail sales in 2025.
OH	Low	Most IOUs achieve savings equal to 50% (First Energy meets 25%) of legislated EEPS targets, rising to 1.0% of retail sales by 2019; stipulated savings reflect a possible reduction in EEPS savings targets, and/or high assumed levels of mercantile customer opt-out, reliance on T&D measures, and noncompliance.
	Medium	IOUs meet EEPS savings targets in 2022 (delayed four years to ramp to 2022) with the exception of First Energy. First Energy meets 50% of its savings targets relying on high assumed levels of mercantile customer opt-out, reliance on T&D measures, and non-compliance.
	High	IOU savings meet legislated EEPS targets of 2% retail sales in 2019, First Energy meets 75% of EEPS target; stipulated savings reflects lower assumed levels (relative to Low/Medium Case) of mercantile customer opt-out, reliance on T&D measures, and non-compliance.
WI	Low	Spending declines from 2.1% of revenues in 2011, back to 2008 level of 1.2%, consistent with legislative rate cap, and remains flat at that level through 2025
	Medium	Same as Low Case through 2015, then spending rebounds to 1.7% of revenues by 2020 (the mid-point between the current spending cap and the maximum historical spending level).

State	Case	Scenario Assumptions
	High	Savings based on lagged 2010 regulatory targets and lagged EEPS policy recommendation in Governor's Task Force on Global Warming, with spending rising to 3.8% of revenues and savings rising to 2.0% of retail sales by 2014.

Table A - 6. Scenario Descriptions for States in the West

State	Case	Scenario Assumptions
AZ	Low	IOUs and Coops meet minimum EERS requirements; annual savings decline over the 2016 to 2020 timeframe because utilities are able to use retroactive credit from pre-EERS programs during that period, and the savings percentage after 2020 remains flat at the 2020 level. SRP savings remain flat at actual 2010 level. Other utilities (munis, tribes, irrigation districts) based on generic method in each scenario.
	Medium	IOUs and Coops are same as the Low Case. SRP savings through 2016 based on targets in 2010 Sustainability Report Summary, and remain flat at the 2016 level thereafter.
	High	IOUs and Coops meet EERS requirements without relying on retroactive credit from historical programs. SRP savings follow the targets in the Sustainable Portfolio Plan through 2020, which rise to 2% of retail sales, and remain constant thereafter (per-unit costs for SRP are higher, to reflect a proportionally lower reliance on M-Power program than in other scenarios).
CA	Low	For IOUs, savings through 2012 are based on level in approved 2010-2012 energy efficiency plans, savings from 2013 to 2014 are based on 90% of levels in the proposed plans, and savings from 2015 onward are equal to 90% of the maximum achievable potential identified in Navigant's 2012 potential study for the CPUC. POU annual savings remain flat at 2010 levels.
	Medium	For IOUs, savings through 2012 are based on level in approved 2010-2012 energy efficiency plans, savings from 2013 to 2014 are based on 100% of levels in the proposed plans, and savings from 2015 onward are equal to 110% of the maximum achievable potential identified in the Navigant potential study. POU annual savings equal 100% of the average annual savings from their 2010- 2020 energy efficiency goals.
	High	For the IOUs, savings through 2012 are based on level in approved 2010-2012 energy efficiency plans, savings from 2013 to 2014 are based on 125% of levels in the proposed plans, and savings from 2015 onward are equal to 130% of the maximum achievable potential identified in the Navigant potential study. POU annual savings equal 125% of the average annual levels identified in their 2010-2020 energy efficiency goals.

Table A - 6. (Continued)

State	Case	Scenario Assumptions
CO	Low	IOUs maintain savings at levels in approved 2011 DSM plans, equal to 0.9% of retail sales in aggregate; Colorado Springs and Fort Collins continue at 2010 savings levels equal to 0.5% of retail sales; generic assumptions for all other POUs are made in each scenario.
	Medium	IOUs meet EEPS targets, which rise to 1.9% of retail sales in 2020; Colorado Springs and Fort Collins achieve and maintain savings of roughly 1% of retail sales by 2015, consistent with CSU's Energy Vision goal.
	High	Same as Medium Case
HI	Low	Statewide savings exhaust, but do not exceed, the 50% energy efficiency allowance under state RPS.
	Medium	Same as Low Case
	High	Same as Low/Medium Cases through 2011, but savings increase from 1.2% of retail sales in 2011 to 2.0% in 2020.
ID	Low	IOU savings are based on projections from each utility's most-recent IRP (Idaho Power's projected savings are well below 2010 program results). POU savings based on achieving their pro-rated share of the NPCC 6th Plan conservation targets, with 60% of the savings assumed to come from utility customer-funded energy efficiency (and the remainder from new codes/standards); savings taper off in the latter years as the retrofit potential is exhausted.
	Medium	Idaho Power savings remain flat at level achieved in 2010, and other IOU savings are based on their IRP projections. POU savings are same as the Low Case through 2020, except that 75% of the conservation target is assumed to come from utility customer-funded energy efficiency, and after 2020, the savings level remains constant at the 2020 level, rather than tapering off as in the Low Case.
	High	IOU savings ramp up from 1.1% of retail sales in 2010 to 1.5% in 2020, and remain flat at that level thereafter. POU savings through 2020 are based on their pro-rated share of NPCC's 6th Plan conservation targets under NPCC's high ramp rate trajectory, and 85% of the savings is assumed to come from utility customer-funded energy efficiency; savings after 2020 remain at the 2020 level, rather than tapering off as under NPCC's projection.

State	Case	Scenario Assumptions
MT	Low	NorthWestern savings based on the targets identified in its 2009 Electric Supply Resource Procurement Plan, which remains flat at approximately 1.0% of retail sales per year. BPA-served POU savings based on same assumptions as Idaho POUs.
	Medium	For NorthWestern, same as Low Case. BPA-served POU savings based on same assumptions as Idaho POUs.
	High	NorthWestern savings ramp up to 1.5% of retail sales per year in 2025. BPA-served POU savings based on same assumptions as Idaho POUs.
NM	Low	IOU savings through 2013 based on most recent approved DSM plans, and savings thereafter decline to the minimum level necessary for EERS compliance, which requires average annual savings of approximately 0.7% over the 2014-2025 period. Savings/spending projections for POUs are based on generic assumptions in each scenario.
	Medium	IOU savings through 2013 based on most recent approved DSM plans, and remain constant at the 2013 level thereafter, equal to 0.8% of retail sales.
	High	IOU savings ramp up to 1.5% of retail sales in 2020
NV	Low	IOU savings and spending through 2013 are based on most recent DSM plans, and thereafter decline to the maximum energy efficiency allowance under the RPS. POU savings/spending projections are based on generic assumptions under each scenario.
	Medium	IOU savings after 2013 remain flat at 2013 levels, equal to 0.7% of retail sales, which far exceeds the energy efficiency allowance under the RPS
	High	Same as Medium Case, except that savings ramp up to 1.5% of retail sales in 2020
OR	Low	PGE and PacifiCorp savings are based on Energy Trust 2010-2014 Strategic Plan savings projection under "current funding" scenario, and remain at the 2014 level thereafter. Idaho Power savings are based on most-recent IRP. POU savings based on same assumptions as in Idaho (ID).
	Medium	PGE and PacifiCorp savings are based on Energy Trust 2010-2014 Strategic Plan savings projection under "IRP-achievable" scenario, and remain at the 2014 level thereafter. Idaho Power savings are based on most-recent IRP. POU savings based on same assumptions as in Idaho (ID).
	High	Statewide savings through 2020 is based on Oregon's pro-rated share of the NPCC's 6th Plan conservation targets, under NPCC's "high ramp rate" trajectory, and 85% of the savings is assumed to come from utility customer-funded energy efficiency; savings after 2020 remain at the 2020 level, rather than tapering off as under NPCC's projection.

Table A - 6. (Continued)

State	Case	Scenario Assumptions
UT	Low	IOU (PacifiCorp) savings remain constant at 2010 level of 0.9% of retail sales. POU savings/spending in all scenarios are based on generic assumptions.
	Medium	PacifiCorp savings follow 2011 IRP schedule, which rises to 1.2% of retail sales in 2021, and stays flat at that level thereafter
	High	PacifiCorp savings rise to 1.5% of retail sales in 2020, and stay flat at that level thereafter
WA	Low	Savings based on achieving WA's pro-rated share of the conservation targets in the NPCC's 6th Plan, with 60% of the savings assumed to come from utility customer-funded energy efficiency (and
		the remainder from new codes/standards); the savings trajectory tapers off in the latter years as the retrofit potential is exhausted.
	Medium	Same as the Low Case, except 75% of the conservation target is assumed to come from utility customer-funded energy efficiency, and after 2020, the savings level remains constant at the 2020 level, rather than tapering off as in the Low Case.
	High	Savings through 2020 are based on WA's pro-rated share of the conservation targets in the NPCC's 6th Plan, under NPCC's "high ramp rate" trajectory, and 85% of the savings is assumed to come from utility customer-funded energy efficiency (and the remainder from new codes/standards); savings after 2020 remain at the 2020 level, rather than tapering off as under NPCC's projection.
WY	Low	PacifiCorp 2009-2013 savings and spending are based on approved DSM plan settlement, and savings from 2014 to 2025 are held constant at 2013 level, equal to 0.5% of retail sales, which is roughly in line with the savings projection from PacifiCorp's 2011 IRP. For all other utilities, spending/savings projections are based on generic assumptions for each scenario.
	Medium	Same as Low Case
	High	Same as the Low Case through 2013, but savings thereafter ramp up to 1.0% of retail sales in 2020 and remain constant at that level thereafter.

Table A - 7. Energy Efficiency Spending Assumed for "Uncommitted" States

Case	Increase in Spending as a Percent of Utility Revenue Relative to 2010														
	2011	2012	2013	2014	2015	2016	2017	2018	2019	2020	2021	2022	2023	2024	2025
Low	0.2%	0.2%	0.2%	0.2%	0.2%	0.2%	0.3%	0.3%	0.3%	0.3%	0.3%	0.3%	0.3%	0.3%	0.3%
Medium	0.4%	0.5%	0.5%	0.5%	0.5%	0.5%	0.5%	0.5%	0.5%	0.5%	0.5%	0.5%	0.5%	0.5%	0.5%
High	0.4%	0.5%	0.5%	0.6%	0.6%	0.7%	0.7%	0.7%	0.8%	0.8%	0.8%	0.8%	0.8%	0.8%	0.8%

A1.3. Cost of Savings Assumptions

Depending on the particular state and scenario, the spending projection may have been estimated from projected first-year savings or vice-versa. In either case, first-year savings were translated into annual spending (or vice-versa) using an assumed cost of savings. We assume that the average cost of savings depends in part on the savings level achieved. To capture this relationship, we developed a generic "cost function" that relates the average cost of first-year electricity savings to the savings level expressed as a percentage of the utility (or state)'s retail sales (see Figure A-2). The y-axis values in the figure are expressed on a normalized (dimensionless) basis, with a cost index of 1.00 at a savings level equal to 1.0% of retail sales. The rationale for this cost function is to reflect the fact that, based on our review of energy efficiency program experience, utility costs to acquire savings (on a dollar-per-MWh basis) can be somewhat higher when portfolio savings levels are low (i.e., annual savings <0.5% of retail sales), due to the effect of fixed program delivery costs and because the utility is implementing pilot programs or is ramping up its administrative and delivery infrastructure. There is also evidence to suggest that program costs increase at relatively high savings targets (i.e., annual savings >1.5% of retail sales) either because rebate levels may be raised in order to achieve higher market penetration or because the utility includes more expensive energy efficiency measures in its program portfolio.

The cost function was then applied to each state by "scaling" the generic cost function shown in Figure A-2 using either state-specific program cost data (if available) or an assumed average cost of savings.[33] State-specific cost of savings data (i.e., cost per first-year MWh saved) were used for 23 states, based on recent program results or recently-approved DSM program plans. For the remaining 27 states, a generic cost of savings value was used to scale the cost function. These states were first categorized as either a low-cost state or a high-cost state. Low-cost states were then assumed to have average program costs equal to $150 per first-year MWh saved at a savings level of 1.0% of retail sales, based on data compiled by ACEEE (Sciortino et al. 2011). High-cost states were assumed to have average program costs equal to $300 per first-yr. MWh saved at a savings level of 1.0% of retail sales, which is based roughly on average costs currently observed among some Northeastern states.

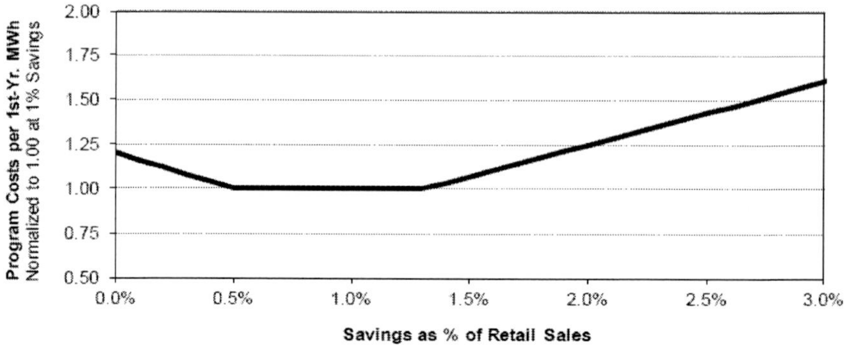

Figure A - 2. Generic Program Cost Function.

A2. Natural Gas Energy Efficiency Spending Projections

We developed low, medium, and high projections of spending through 2025 on energy efficiency programs funded by customers of natural gas utilities. Given that spending on natural gas programs represents a relatively small portion of total (electric plus gas) customer-funded spending, we used a simpler and more standardized approach to project future spending, compared to electric energy efficiency programs.

A2.1. Revenue Projections

Projections of revenue from retail natural gas sales to residential, commercial, and industrial customers (i.e., excluding sales to electric utilities) were developed in a similar manner as the baseline projections of revenue from retail electricity sales. Retail sales and retail price projections were first developed for each state by applying annual growth rate projections from the AEO2012 reference case forecast to actual 2010 retail sales and price data for each state, as reported by EIA. Retail gas sales include sales to residential, commercial, industrial, and transportation sectors, but exclude sales to the electric power sector. Average annual retail gas prices were calculated as the average of EIA's forecast of prices for the residential, commercial, industrial, and transportation sectors, weighted by the quantity of sales to each sector. The natural gas retail sales and retail price projections in AEO2012 are specified at the census-region level. Thus, the growth rates in each census region were applied to each state in their respective region. Revenue projections were calculated by multiplying projected retail gas prices by projected retail gas sales, and were converted to nominal dollars using the AEO2012 reference case forecast of the GDP chain-type price index. Unlike the electricity revenue projections, no adjustments were made to the natural gas revenue projections to account for differing levels of energy efficiency savings across scenarios.

A2.2. Scenario Definitions

For the purpose of developing projections of utility customer funding of gas efficiency programs, we first grouped states into three categories: Tier I consists of the 13 states that comprise more than 80% of current national funding for gas efficiency programs, Tier II consists of another 15 states where 2010 spending on gas efficiency programs exceeded $0.50 per capita, and Tier III consists of the remaining 23 states. Table A-9 identifies the states in each group.

Table A - 8. Analysis Framework for Natural Gas Energy Efficiency Programs

Tier I	CA
Tier II	AR
Tier III	AK, AL, AZ, DC, DE, FL, GA, HI, KS, KY, LA, ME, MS, MT, NC, ND, NE, OK, SC, TN, TX, VA, WV

Tier I Scenario Definitions

For Tier I states, gas efficiency program spending projections are based on state-specific policies, gas DSM program plans, and regulatory decisions that set savings targets for gas utilities. Table A-10 describes the specific assumptions underlying the three scenarios in each Tier I state. For most Tier I states, the low and medium case spending projections track the most recent multi-year gas DSM program plans to their terminal year (typically 2012 to 2014). In the low case, we assume that spending on residential gas efficiency programs in most Tier I states will decline by 50% to 75% relative to the level in the terminal year of the most recent DSM plan, while spending on commercial and industrial (C&I) programs will decline by roughly 20%. This decline in spending is due to the combination of sustained low natural gas prices, which reduce the cost effectiveness of gas efficiency programs and tightening federal minimum efficiency standards for gas furnaces, which reduce the savings for voluntary programs. In the medium case, we assume a more modest drop-off in residential program spending, typically to 50% of the level from the terminal year of the most recent gas DSM Plan, but that C&I program spending increases slightly, as program managers shift spending towards markets with greater savings opportunities. In both the low and medium scenario, we assume that gas efficiency programs for low-income programs, however, remain constant at the level from the last year of the DSM plan, as these programs meet broader policy objectives (e.g. equity, reductions in bill arrearages), and therefore are less susceptible to the dynamics putting downward pressure on gas program spending for the other sectors. Finally, in the high case, we assume that many Tier I states achieve gas savings levels on par with gas EERS targets in a number of states (i.e., 1.0- 1.5% of total utility retail sales).

Table A - 9. Gas Energy Efficiency Program Scenario Descriptions: Tier I States

<table>
<tr><th>State</th><th>Case</th><th>Scenario Description</th></tr>
<tr><td rowspan="3">CA</td><td>Low</td><td>Program savings are assumed to track 60% of the total gross market potential excluding codes and standards, based on the Navigant (2012) potential study; spending is calculated from savings, based on the planned savings and spending in the IOU 2013-2014 program plans.</td></tr>
<tr><td>Medium</td><td>Similar to Low Case, except savings are assumed to track 80% of the total gross market potential excluding codes and standards.</td></tr>
<tr><td>High</td><td>Spending as a percent of revenues remains flat at 2012 levels</td></tr>
<tr><td rowspan="2">CO</td><td>Low</td><td>Spending tracks most recent gas DSM plan to terminal year. Thereafter, residential program spending as a percentage of revenues declines over time by 75% and C/I spending declines by 20%, to reflect impact of furnace standards and low gas prices. Low-income program spending as a percentage of revenues continues at the same level as in the DSM plan.</td></tr>
<tr><td>Medium</td><td>Spending tracks most recent gas DSM plan to terminal year. Thereafter, residential program spending as a percentage of revenues declines over time by 50%, while C/I spending increases 30%, as funds are shifted across market segments. Low-income program spending as a percentage of revenues continues at the same level as in the DSM plan.</td></tr>
</table>

State	Case	Scenario Description
	High	Savings ramps up to 1% of retail sales to non-transportation gas customers
IA	Low	Spending tracks most recent gas DSM plan to terminal year. Thereafter, residential program spending as a percentage of revenues declines over time, beginning in 2017, by 75% and C/I spending declines by 20%, to reflect impact of furnace standards and low gas prices. Low-income program spending as a percentage of revenues continues at the same level as in the DSM plan.
	Medium	Spending tracks most recent gas DSM plan to terminal year. Thereafter, residential program spending as a percentage of revenues declines over time by 50%, while C/I spending increases 30%, as funds are shifted across market segments. Low-income program spending as a percentage of revenues continues at the same level as in the DSM plan.
	High	Savings ramps up to 1.5% of retail sales to non-transportation gas customers, on par with the gas EERS targets in IL and MN
IL	Low	See Iowa (IA) Low Case
	Medium	See Iowa (IA) Medium Case
	High	Savings ramps up to meet Illinois gas EERS savings targets (1.5% of retail sales to non-transportation gas customers)
IN	Low	See Iowa (IA) Low Case
	Medium	See Iowa (IA) Low Case
	High	Savings ramps up to 1% of retail sales to non-transportation gas customers
MA	Low	Spending tracks most recent gas DSM plan to terminal year. Thereafter, residential program spending as a percentage of revenues declines over time by 50% and C/I spending declines by 20%, to reflect impact of furnace standards and low gas prices. Low-income program spending as a percentage of revenues continues at the same level as in the DSM plan.
	Medium	Spending tracks most recent gas DSM plan to terminal year. Thereafter, residential program spending as a percentage of revenues declines over time by 20%, while C/I spending increases 30%, as funds are shifted across market segments. Low-income program spending as a percentage of revenues continues at the same level as in the DSM plan.
	High	Savings ramps up to 1.5% of total statewide gas retail sales, a proxy for achievement of all cost-effective gas energy efficiency
MI	Low	See Iowa (IA) Low Case
	Medium	See Iowa (IA) Medium Case
	High	See Iowa (IA) High Case
MN	Low	See Iowa (IA) Low Case
	Medium	See Iowa (IA) Medium Case

Table A - 9. (Continued)

State	Case	Scenario Description
	High	Savings ramps up to meet Minnesota gas EERS savings targets (1.5% of retail sales to non-transportation gas customers)
NJ	Low	Similar to Colorado (CO) Low Case, except total spending is de-rated by 70% to reflect historical gap between actual expenditures and planned budgets.
	Medium	Similar to Colorado (CO) Medium Case, except total spending is de-rated by 70% to reflect historical gap between actual expenditures and planned budgets, and C/I spending as a percentage of revenues remains flat, rather than increasing.
	High	Savings ramps up to 1% of total statewide retail sales
NY	Low	Similar to Massachusetts (MA) Low Case, but projections to 2015 are based on achievement of New York's gas EERS targets
	Medium	Similar to Massachusetts (MA) Medium Case, but projections to 2015 are based on achievement of New York's gas EERS targets
	High	Similar to New Jersey (NJ) High Case, but projections to 2015 are based on achievement of New York's gas EERS targets
OR	Low	See Iowa (IA) Low Case
	Medium	See Iowa (IA) Medium Case
	High	See Iowa (IA) High Case
UT	Low	See Iowa (IA) Low Case
	Medium	Similar to Colorado (CO) Medium Case, except residential program spending declines by 70% rather than 50%, based on information provided by program managers.
	High	See Iowa (IA) High Case
WI	Low	Spending is based on the current legislative cap (1.2% of revenues)
	Medium	Same as Low Case
	High	See Iowa (IA) High Case

Tier II Scenario Definitions

The 15 Tier II states have relatively aggressive spending levels on a per capita basis, but small populations and therefore small spending levels in absolute terms. Thus, for simplicity, the spending projections for these states were developed based on regional benchmark trajectories that were developed from the projections for Tier I states in the corresponding region. These regional benchmark trajectories were developed by averaging the change, relative to 2010, in spending as a percentage of gas distribution utility revenues across Tier I states in each census region (see Table A-11). Those growth curves were then applied to the 2010 spending for each Tier II state.

Table A - 10. Gas Energy Efficiency Program Spending: Tier II States

Case		Increase (Relative to 2010) in Customer-Funded Gas EE Spending as a Percent of Revenues from Retail Gas Sales															
		2010	2011	2012	2013	2014	2015	2016	2017	2018	2019	2020	2021	2022	2023	2024	2025
Northeast	Low	0.0%	0.8%	1.3%	1.4%	1.4%	1.4%	1.4%	0.9%	0.5%	0.5%	0.5%	0.5%	0.5%	0.5%	0.5%	0.5%
	Medium	0.0%	0.8%	1.3%	1.4%	1.4%	1.4%	1.4%	1.1%	1.1%	1.1%	1.2%	1.2%	1.2%	1.2%	1.2%	1.2%
	High	0.0%	0.8%	1.3%	1.6%	1.8%	2.0%	2.3%	2.7%	2.9%	3.3%	3.6%	3.4%	3.3%	3.3%	3.2%	3.2%
Midwest & South	Low	0.0%	0.2%	0.4%	0.6%	0.6%	0.6%	0.6%	0.3%	0.0%	0.0%	0.0%	-0.2%	-0.2%	-0.2%	-0.2%	-0.2%
	Medium	0.0%	0.2%	0.4%	0.6%	0.6%	0.6%	0.6%	0.6%	0.6%	0.6%	0.4%	0.2%	0.2%	0.2%	0.2%	0.2%
	High	0.0%	0.2%	0.4%	0.6%	0.8%	0.8%	1.0%	1.2%	1.3%	1.5%	1.5%	1.4%	1.3%	1.3%	1.3%	1.2%
West	Low	0.0%	0.1%	0.2%	0.1%	0.3%	0.3%	0.3%	-0.2%	-0.6%	-0.6%	-0.7%	-1.0%	-1.0%	-1.0%	-1.0%	-1.0%
	Medium	0.0%	0.1%	0.2%	0.1%	0.3%	0.3%	0.3%	0.2%	0.2%	0.2%	-0.1%	-0.5%	-0.5%	-0.6%	-0.7%	-0.7%
	High	0.0%	0.1%	0.2%	0.0%	0.4%	0.6%	0.9%	1.1%	1.2%	1.4%	1.5%	1.4%	1.3%	1.2%	1.2%	1.1%

As an example, in the medium case, spending for the three Tier I states in the northeast (MA, NY, and NJ) is projected to increase by, on average, 0.6% of revenues; thus, the same 0.6% increase in spending as a percent of revenues was stipulated for the Tier II northeastern states in the medium case.

Tier III Scenario Definitions

For states in Tier III that currently have little or no customer-funded gas efficiency program activity, we assumed that spending will stay flat as an absolute dollar value in the low scenario. In the medium scenario, we project that gas efficiency programs will continue spending at their present percentage of retail gas revenues. In the high case, we assume that gas efficiency programs will increase program spending to approximately 0.25% of revenues above their current levels (e.g., if current gas program spending is equal to 0.5% of revenues, then we assume in the high case that spending rises to 0.75% of revenues).

Technical Appendix B: State-Level Spending and Savings Projections

The results presented in Section 4 focused on regional and national spending and savings projections. This technical appendix provides additional state-level details. Table B - 1 presents projected electric efficiency program spending by state, Table B - 2 presents projected electric efficiency program savings by state, and Table B - 3 presents projected gas efficiency program spending by state. Note that the spending projections are presented in terms of nominal dollars, as used throughout the report, and the savings projections are presented in terms of first-year GWh. Please refer to Technical Appendix A for details on the underlying scenario definitions and assumptions used for each state in order to develop these spending and savings projections.

Table B - 1. Electric Energy Efficiency Program Spending Projections by State ($M, nominal

State	2010*		Low			Medium			High	
		2015	2020	2025	2015	2020	2025	2015	2020	2025
AK	-	2	3	4	5	5	6	6	9	9
AL	9	23	34	38	51	61	68	60	94	104
AR	9	38	44	48	46	59	65	48	73	85
AZ	88	191	117	136	199	138	160	250	321	361

State	2010*		Low			Medium			High	
		2015	2020	2025	2015	2020	2025	2015	2020	2025
CA	938	755	721	701	927	878	852	1,203	1,050	1,021
CO	48	62	71	82	105	167	180	107	173	186
CT	108	86	88	89	102	109	123	278	282	301
DC	15	8	17	18	19	20	22	19	24	26
DE	7	0	7	7	7	7	7	12	21	22
FL	165	145	134	160	234	278	325	310	433	727
GA	15	60	73	83	70	93	106	84	143	216
HI	25	66	78	95	66	78	95	62	100	106
IA	59	103	110	119	123	130	140	161	264	288
ID	57	32	40	34	43	51	57	48	71	77
IL	85	233	235	235	314	373	389	666	675	681
IN	5	150	255	267	200	385	395	202	389	399
KS	6	8	13	14	19	21	24	23	34	37
KY	21	45	54	63	75	89	102	93	124	142
LA	-	17	28	32	39	45	53	48	72	84
MA	245	200	221	243	568	580	593	613	625	666
MD	74	172	182	193	174	181	192	168	269	292
ME	14	22	24	27	40	42	46	48	50	54
MI	75	167	172	185	167	172	185	205	278	338
MN	107	110	117	129	121	141	153	175	284	296
MO	36	91	87	93	97	184	188	135	322	342
MS	6	16	22	26	34	42	48	39	63	72
MT	17	14	17	17	17	19	21	20	27	33
NC	52	66	113	124	103	116	147	110	202	314
ND	8	2	3	4	5	5	6	6	9	10
NE	-	5	8	9	12	13	15	14	21	23
NH	18	20	22	26	30	48	52	41	98	101
NJ	191	129	93	97	184	188	194	211	348	532
NM	18	22	25	27	27	31	36	35	59	66
NV	33	16	18	21	36	41	48	49	97	109
NY	482	552	559	608	778	653	704	778	865	913
OH	70	78	162	175	170	297	451	183	463	484
OK	21	53	59	67	57	111	123	58	121	195
OR	135	107	126	131	158	181	199	152	223	243
PA	98	210	224	245	240	240	240	263	329	405
RI	44	43	46	51	73	74	74	96	97	103
SC	15	75	89	103	89	114	131	109	140	210
SD	0	2	3	4	5	5	6	6	8	9
TN	38	47	58	66	93	121	136	98	172	189
TX	114	119	150	171	163	215	254	246	252	285
UT	48	54	60	70	35	46	51	39	63	71
VA	0	27	34	39	43	63	72	115	130	143
VT	36	44	45	48	44	45	48	64	63	66
WA	218	167	229	192	209	287	317	295	428	466
WI	74	83	92	99	83	126	135	255	264	268
WV	-	6	9	10	13	15	17	16	24	27

Table B - 1. (Continued)

State	2010*		Low			Medium			High	
		2015	2020	2025	2015	2020	2025	2015	2020	2025
WY	3	11	12	12	12	13	14	14	23	25
Total	3,948	4,757	5,202	5,536	6,522	7,398	8,062	8,336	10,769	12,222

* Source of 2010 spending data: CEE (2012), with minor modifications.

Table B - 2. Electric Energy Efficiency Program Savings Projections by State (First-Year GWh

State	2009*		Low			Medium			High	
		2015	2020	2025	2015	2020	2025	2015	2020	2025
AK	1	11	16	16	29	28	27	36	46	45
AL	63	121	164	170	288	314	322	337	496	520
AR	60	224	233	236	268	313	318	280	396	421
AZ	571	1,101	761	810	1,164	894	945	1,281	1,430	1,474
CA	2,293	2,497	2,202	1,979	3,067	2,681	2,402	3,670	3,214	2,887
CO	255	327	346	365	546	696	704	556	725	736
CT	250	287	268	248	338	332	342	633	591	577
DC	56	44	90	89	113	111	110	113	133	130
DE	0	0	35	32	40	35	32	73	112	110
FL	365	384	324	343	640	697	747	877	1,166	1,796
GA	54	336	370	388	399	499	517	493	800	1,113
HI	113	176	185	195	176	185	195	172	218	211
IA	410	422	415	410	503	489	480	653	861	864
ID	186	193	218	167	254	274	282	283	351	352
IL	553	1,085	1,005	923	1,465	1,600	1,528	2,494	2,331	2,156
IN	40	885	1,289	1,239	1,183	1,675	1,574	1,191	1,697	1,596
KS	1	42	60	62	103	106	108	126	175	177
KY	65	253	278	294	439	476	501	543	666	699
LA	0	87	129	139	205	220	236	261	377	401
MA	459	563	572	578	1,077	1,008	944	1,105	1,037	1,013
MD	274	688	666	648	698	666	648	673	928	891
ME	94	114	113	117	183	175	171	215	204	202
MI	376	956	936	919	990	936	919	959	1,191	1,279
MN	638	651	635	640	714	766	760	967	1,232	1,178
MO	86	348	355	349	385	755	707	611	1,170	1,086
MS	31	84	111	118	193	220	231	219	341	358
MT	57	75	82	77	88	93	93	101	124	135
NC	52	364	591	591	608	631	733	650	1,095	1,562
ND	3	11	15	16	25	26	27	31	43	44
NE	65	26	37	38	62	64	65	75	105	107
NH	68	60	60	64	90	131	128	123	214	201
NJ	497	211	112	110	311	197	191	374	631	1,018
NM	59	123	125	126	148	156	165	197	282	289
NV	439	89	92	93	211	223	239	287	491	507

State	2009*		Low			Medium			High	
		2015	2020	2025	2015	2020	2025	2015	2020	2025
NY	950	2,563	2,311	2,299	2,647	2,414	2,449	2,647	2,894	2,936
OH	530	458	875	864	1,007	1,522	1,857	1,084	2,082	1,999
OK	20	174	188	197	198	368	379	208	415	606
OR	292	472	507	484	642	676	681	640	785	782
PA	279	831	815	815	951	872	798	1,042	1,197	1,346
RI	82	107	103	102	149	140	131	172	159	155
SC	46	311	341	362	383	447	470	469	562	760
SD	22	10	15	15	24	25	26	30	42	44
TN	121	256	292	305	552	658	674	581	933	942
TX	751	650	755	786	924	1,118	1,211	1,451	1,364	1,409
UT	177	210	220	232	233	281	286	268	371	381
VA	1	189	214	223	318	454	479	972	1,000	1,003
VT	90	114	110	105	114	110	105	176	164	164
WA	665	722	909	698	903	1,137	1,152	1,231	1,487	1,482
WI	584	435	441	435	435	606	592	1,301	1,220	1,161
WV	0	30	42	44	70	72	76	89	124	128
WY	7	62	62	61	70	68	66	81	122	124
Total	13,147	20,433	21,092	20,613	26,624	28,641	28,823	33,102	39,792	41,561

* Source of 2009 savings data: Sciortino et al. (2011).

Table B - 3. Gas Efficiency Program Spending Projections by State ($M, nominal

State	2010*		Low			Medium			High	
		2015	2020	2025	2015	2020	2025	2015	2020	2025
AK	0	0	0	0	0	0	0	0	1	1
AL	0	0	0	0	0	0	0	2	4	7
AR	2	11	3	2	11	10	8	14	29	30
AZ	2	2	2	2	2	3	4	3	5	8
CA	201	152	100	118	171	134	157	198	249	315
CO	19	15	10	12	15	17	17	31	74	83
CT	12	31	23	27	31	35	41	39	73	80
DC	0	0	0	0	0	0	0	0	1	2
DE	0	0	0	0	0	0	0	0	1	2
FL	11	11	11	11	12	14	17	13	17	23
GA	0	0	0	0	0	0	0	3	8	14
HI	0	0	0	0	0	0	0	0	0	0
IA	40	51	39	34	51	55	55	66	115	127
ID	2	3	1	1	4	4	1	4	12	13
IL	26	126	86	69	126	127	118	126	235	254
IN	11	18	14	12	18	18	18	18	45	44
KS	0	0	0	0	0	0	0	1	3	6
KY	1	1	1	1	1	1	1	2	4	7
LA	0	0	0	0	0	0	0	4	10	18
MA	72	139	114	136	139	157	186	182	277	308

Table B - 3. (Continued)

State	2010*	Low			Medium			High		
		2015	2020	2025	2015	2020	2025	2015	2020	2025
MD	6	18	8	3	18	18	15	23	45	47
ME	1	1	1	1	1	1	2	1	2	3
MI	41	87	74	74	87	95	99	99	138	149
MN	36	40	33	33	40	48	52	47	63	69
MO	7	20	9	4	20	21	17	26	52	55
MS	0	0	0	0	0	0	0	1	2	4
MT	0	0	0	0	0	0	0	1	1	2
NC	1	1	1	1	1	2	2	3	6	11
ND	0	0	0	0	0	0	0	0	1	1
NE	0	0	0	0	0	0	0	1	2	4
NH	4	15	15	18	15	18	21	17	26	30
NJ	126	101	78	72	101	83	99	156	247	273
NM	2	4	1	1	5	5	1	5	15	16
NV	3	6	1	1	6	7	1	7	21	23
NY	39	158	135	160	158	178	211	158	371	410
OH	32	68	38	24	68	69	61	84	152	156
OK	0	0	0	0	0	0	0	2	6	10
OR	23	40	32	28	40	45	45	41	62	70
PA	13	95	53	63	95	103	122	132	266	287
RI	5	13	10	12	13	15	17	17	31	34
SC	0	0	0	0	0	0	1	1	3	5
SD	1	4	2	1	4	4	3	5	9	9
TN	0	0	0	0	0	0	0	2	4	7
TX	2	2	2	2	2	3	3	10	22	39
UT	36	34	23	16	34	34	20	36	65	72
VA	3	3	3	3	3	4	5	5	9	14
VT	2	4	3	4	4	4	5	4	7	8
WA	29	14	3	3	15	17	3	18	49	53
WI	24	21	25	30	21	25	30	104	122	132
WV	0	0	0	0	0	0	0	1	1	3
WY	0	1	0	1	2	1	1	2	7	8
Total	838	1,313	957	982	1,335	1,376	1,458	1,721	2,968	3,346

* Source of 2010 spending data: CEE (2012), with minor modifications.

End Notes

[1] The American Recovery and Reinvestment Act (ARRA) provided a massive but temporary infusion of federal funding for energy efficiency (~$15-20B in programs administered by federal, state and local governments to be spent over three years) (Goldman et al. 2011).

[2] Geller et al. (2006) provide an overview of the efficiency policy landscape among nations in the Organization of Economic Cooperation and Development, including the U.S., while Gillingham et al. (2006) provide a comprehensive review for the United States.

[3] Energy efficiency programs administered by U.S. gas distribution utilities have also increased over time but are much smaller in size than electric efficiency programs (York et al. 2012).

[4] By virtue of limiting the analysis to current energy efficiency policies, we do not consider the potential impact of major new federal (or state) policy initiatives (e.g., a national energy efficiency resource standard, clean energy standard, or carbon policy) that could result in customer-funded energy efficiency program spending and savings that exceed the values in our High Case.

[5] In this study, we define Energy Efficiency Resource or Portfolio Standards as requirements under statute or regulatory order that some or all utilities within a state (e.g., all utilities or investor-owned utilities only) achieve specified minimum savings levels over a period greater than three years. States with shorter term DSM plans (i.e. one to three years) and/or multi-year efficiency budgets approved by state regulators are separately listed. Note that other entities (e.g., ACEEE) that track the status of energy efficiency policies in various states may use slightly different criteria for defining an EERS than LBNL; thus their tallies of the number of states with such policies may differ.

[6] In Arkansas, the process began with a collaborative among regulators, utilities and other stakeholders, then proceeded to "quick start" programs designed to test the viability of utility customer-funded programs in that jurisdiction and begin building program infrastructure. In the final step, regulators set modest but increasing savings targets.

[7] For example, Duke Energy Carolinas and Progress Energy Carolinas are subject to an RPS in North Carolina in which energy efficiency is an eligible resource, and both submitted a *pro rata* version of the same efficiency plan from North Carolina for the rest of their service territory in South Carolina. Likewise, West Virginia's requirement that an American Electric Power subsidiary initiate efficiency programs resulted in submission of similar program plans in neighboring Virginia.

[8] The Tennessee Valley Authority (TVA) is the largest U.S. public power utility and serves 155 distributors and 57 industrial customers in TN, KY, AL, MS, GA, NC and VA.

[9] The systems benefit charge in California expired at the end of 2011, although legislative efforts are underway to renew it.

[10] Electric utility expenditures on load management programs in 2010 represented an additional $0.9 billion (CEE 2012).

[11] Metrics based on total budget for energy efficiency tend to favor states with large populations. It is important to note that program administrators in several small states (e.g. VT, RI, and IA) have significant energy efficiency budgets, if metrics are based on efficiency spending per capita. In the U.S., the 10 leading states spend more than $25 per capita on utility customer-funded electric efficiency programs, while the average spending on these programs among the 50 states is about $12 per capita.

[12] See approach section on modeling of gas programs for more details.

[13] On a national basis, electric energy efficiency spending in 2010 was allocated among market sectors as follows: commercial and industrial (47%), residential (28%), low-income residential (8%) and other programs or expenditure categories not directly attributable to a sector (16%).

[14] In the context of this report, "spending" refers to the flow of money from the energy efficiency program administrator into the market, including all program administration costs but excluding performance incentives. To the extent possible, electric spending projections are intended to reflect "gross" savings, prior to accounting for free-riders or free-drivers. This approach was taken in order to abstract from potential inconsistencies across states in methods for estimating net-to-gross ratios. However, the underlying data used to derive the

cost of savings for some states were not explicit about whether savings are reported in "net" or "gross" terms; thus, some ambiguity exists in whether the spending projections for a number of states reflect net or gross savings. Gas efficiency program savings projections were not included for several reasons (e.g., relative paucity of mature, multi-year gas efficiency programs from which to draw reliable data).

[15] These seven uncommitted states include: AK, KS, LA, ND, NE, SD, and WV.

[16] We do not present projections of natural gas program savings, as insufficient data exists to link the projected spending amounts to specific savings trajectories.

[17] Note that energy savings number cited here represents first-year savings from programs implemented in 2009, and should not be confused or compared with other estimates (e.g., CEE 2012) that refer to the combined impact in any given year from both programs implemented in that year and from programs implemented in prior years.

[18] To the extent possible, spending projections are intended to reflect "gross" savings (i.e., prior to accounting for free-riders or spillover effects). This approach was taken in order to abstract from potential inconsistencies across states in methods for estimating net-to-gross ratios.

[19] For many states, our analysis assumes constant savings percentages from 2020 to 2025; those assumptions are reflected in the national totals in Figure 7, which similarly shows a flat or slight decline in savings percentages from 2020 to 2025.

[20] EIA's National Energy Modeling System (NEMS) is calibrated to historical data on end-use stock efficiency and shipments, and the customer adoption simulation assumes, in essence, that consumers will continue purchasing equipment that exceeds minimum efficiency standards to the same extent as has historically occurred. This estimate of incremental savings from efficiency programs is based on ACEEE data for national electric efficiency program savings for 2006-2010, and savings for 2000-2005 are estimated based electric efficiency program spending for those years.

[21] One must interpret this finding with a certain degree of caution given that: (a) EIA's 2012 reference case load forecast projects much slower growth in electricity demand and in economic activity than has historically occurred, and (b) uncertainty regarding the precise extent to which EIA's load forecast accounts for the impacts of future electric utility customer-funded efficiency programs. Our results suggest that additional analysis of the amount of future energy efficiency program savings that is implicit in EIA's reference case forecast and more consistent accounting of free rider and spillover effects is warranted, although beyond the scope of this study.

[22] Other aspects of the broader market and policy context that may impact future customer-funded energy efficiency program activity include the development of alternative utility business models, increasing capital costs for conventional generation technologies, greenhouse gas mitigation policies, and the lasting effects of ARRAfunding on energy efficiency program delivery infrastructure and energy efficiency potential.

[23] Lawmakers in Wisconsin and regulators in Florida, for example, have both cited rate impacts in repealing or lowering energy savings targets.

[24] Actual diversions of SBC funds to state general funds have been considered by state legislators in a number of states and have actually occurred in several states. For example, nearly a third of program revenues in Connecticut were redirected to the state general fund and California lawmakers considered diverting $161 million in gas SBC funds to the state general fund but a state court denied the transfer.

[25] The trajectory for gas prices, and the implications for spending and performance of gas energy efficiency programs, could change in response to, for example, tighter regulation of

hydraulic fracturing, an acceleration in the expected increase of demand among gas-fired generators or a rapid increase in exports of liquified natural gas.

[26] Annual fuel use efficiency (AFUE) is an equipment rating intended to measure the season-long, average efficiency of equipment as a ratio of thermal energy output to fuel energy input. An AFUE of 78%, the current U.S. standard for furnaces, represents an average of 78 Btus of heat for every 100 Btus of energy in the combusted fuel.

[27] Efforts to limit these emissions span multiple sets of regulations – for air toxics, for nitrogen and sulfur oxides, for greenhouse gases, for managing coal ash and for limiting once-through cooling for generators – and each of these regulations has its own timeline and likelihood of coming into effect.

[28] Coal-fired generators are the nation's largest single source of acid gases, carbon dioxide and air toxics such as mercury. The oldest coal-fired generators in the U.S., generally those of 1960s vintage or earlier would be most affected by these environmental regulations. See CRS (2011), Brattle Group (2010) and Bipartisan Policy Commission (2011) for a more detailed discussion of these regulations, their timing and the projected impacts on the electric power industry.

[29] For example, Michigan and Illinois have spending caps in their EERS legislation. In Wisconsin, lawmakers rescinded regulatory discretion over program spending and capped spending at about half the levels anticipated to meet original savings targets. In Florida, the PSC continues to rely heavily on the RIM test to screen energy efficiency programs; the RIM test highlights potential rate impacts on non-participants rather than reductions in average customer bills from cost-effective efficiency investments.

[30] Related, large electricity customers in many utility service areas may either "opt out" of paying charges for energy efficiency programs or direct most or all of their share of those charges into their own, "self-direct" energy efficiency investments.

[31] NEMS is calibrated to historical data on end-use stock efficiency and shipments, and the customer adoption simulation operates under the assumption that, in essence, consumers will continue purchasing equipment that exceeds minimum efficiency standards to the same extent as has historically occurred.

[32] These seven uncommitted states include: AK, KS, LA, ND, NE, SD, and WV

[33] For example, if data for a given state indicate that average program costs are $200 per first-yr. MWh saved at savings equal to 1.0% of retail sales, then the generic cost function would yield an average cost of $250 per first-yr. MWh at savings equal to 2.0% of retail sales (i.e., 1.25 times the cost at a savings level equal to 1.0% of retail sales).

In: Utility Customer-Funded Energy ... ISBN: 978-1-62948-939-1
Editor: Trevor Maselli

Chapter 2

THE SHIFTING LANDSCAPE OF RATEPAYER-FUNDED ENERGY EFFICIENCY IN THE U.S.*

Galen Barbose, Charles Goldman and Jeff Schlegel

EXECUTIVE SUMMARY

Over the last two decades, utility ratepayer funding for energy efficiency programs has seen both booms and busts. Currently, roughly 35 states implement some set of ratepayer-funded electric and/or natural gas energy efficiency programs, with a total U.S. budget of $3.1 billion in 2008 (CEE 2008).[1] The top 10 states account for about 80% of this total, each of which planned to spend $100 million or more in 2008. California, the undisputed heavyweight in terms of the absolute magnitude of its spending on energy efficiency, represented one-third of the total U.S. energy efficiency program budget in 2008.

A proliferation of new state-level policies enacted over the past several years suggests that the next decade may see a dramatic and sustained increase in overall funding levels, and a fundamental re-drawing of the energy efficiency map. These new policies include: energy efficiency portfolio or resource standards (EEPS or EERS), requirements that utilities acquire all

* This is an edited, reformatted and augmented version of the Office of Electricity Delivery and Energy Reliability and by the Office of Energy Efficiency and Renewable Energy, Weatherization and Intergovernmental Program of the U.S. Department of Energy, DE-AC02-05CH11231, dated October, 2009.

cost-effective energy efficiency, strengthened integrated resource planning (IRP) or demand-side management (DSM) planning requirements, and regulatory incentive mechanisms to better align utility financial interests with improvements in customer energy efficiency.

To assess the impact of these new policies on energy efficiency spending and savings, we developed a set of projections (low, medium, and high) of ratepayer-funded energy efficiency program spending and savings through 2020, based on a state-by-state review of energy efficiency policies currently on the books or in the pipeline, as well as recent IRPs and DSM plans. The projections suggest the following set of trends:

- Total ratepayer-funded energy efficiency program spending in the U.S. (90% of which is targeted to electric end-uses) is projected to increase from $3.1 billion in 2008 to $5.4-$12.4 billion per year in 2020, with a Medium Case projection of $7.5 billion.
- Much of the projected increase will be centered in populous states that, historically, have been relatively minor players on the national energy efficiency stage, but have recently enacted aggressive new energy efficiency policies, including: Illinois, Maryland, Michigan, North Carolina, Ohio, and Pennsylvania. These six states, which together represented less than 4% of energy efficiency program spending in 2008, account for more than 60% of the projected increase in total U.S. spending from 2008-2020 under our Low Case (and smaller percentages under the other cases).
- Among states that have traditionally provided strong support for ratepayer-funded energy efficiency programs, significant funding increases (i.e., >$200 million per year) are projected under both the Mid Case and High Case scenarios for New Jersey, New York, and Massachusetts. In the High Case, significant funding increases are also projected for a number of other traditional leaders, including California, Connecticut, Minnesota, and Wisconsin.
- Across all three scenarios, spending levels become significantly more evenly distributed across states than is currently the case. In California, ratepayer funding is projected to decline or remain relatively flat, and the gap between it and other leading states narrows considerably, with all of the top-10 states (in terms of absolute funding levels) reaching at least $200 million per year in the Low Case and at least $400 million per year in the High Case. Program funding also becomes considerably more evenly distributed at a

national level, with a much greater percentage of future energy efficiency spending projected to occur outside of the top-10 states than is currently the case (i.e., roughly 40% by 2020 compared to 20% today).

- Projected increases in ratepayer funding for electric energy efficiency programs are estimated to yield annual electric energy savings in 2020 ranging from 0.45% (Low Case) to 0.93% (High Case) of total U.S. retail sales in that year, with a Mid Case estimate of 0.58%. This compares to estimated annual savings of 0.34% of U.S. retail sales in 2008.
- Cumulative electricity savings projected from ratepayer-funded programs implemented over the 2010-2020 period equal 4.7% of EIA's reference case forecast of 2020 retail electricity sales in the Low Case, 6.1% in the Medium Case, and 8.6% in the High Case.
- Energy savings from ratepayer-funded efficiency programs, as a result of the state-level policies examined in this report, have important implications for the potential incremental impact (in terms of both its size and distribution) of a national EERS or clean energy standard. For example, an EERS requiring *cumulative* savings of only 5% of retail sales by 2020 would result in little or no incremental increase in energy efficiency savings than would likely occur in the absence of such a policy.

The Medium and High Case spending projections represent more than a doubling and quadrupling, respectively, of current national energy efficiency program funding levels. And in some states, program funding is projected to increase by an order of magnitude or more under one or more of the three scenarios. While achieving a ramp-up of this scale over the next decade is most likely feasible, states and program administrators are expected to face a number of near- term and longer-term challenges. These include:

- The economic downturn, which may affect both the ability of programs to acquire savings, and the political feasibility of increasing ratepayer funding for energy efficiency programs;
- General aversion by public utility commissions to the *short-term* rate impacts associated with large-scale energy efficiency implementation (a longer-term issue distinct from the economic downturn);

- Coordination with state/federal energy efficiency programs, including, in the near-term, programs funded through *The American Recovery and Reinvestment Act* (aka, the "stimulus bill");
- The need to develop innovative program designs to reach deeper and broader savings, in order to achieve statewide savings goals significantly beyond what is currently being achieved;
- The effect of new state and/or Federal appliance and lighting efficiency standards on the remaining market potential that can be captured by voluntary energy efficiency programs;
- The need to develop the institutional framework for effective regulatory oversight of ratepayer-funded energy efficiency programs in states that historically have not had significant program activity; and
- The potential, most likely near-term, shortage of trained personnel in the energy efficiency services sector.

1. INTRODUCTION

Over the last two decades, utility ratepayer funding for energy efficiency programs – and the associated energy savings – has seen both booms and busts. Currently, about 35 states implement ratepayer-funded energy efficiency programs, with a total U.S. budget of $3.1 billion in 2008, approximately 80% of which is concentrated in just ten states (CEE 2008).[2] However, a proliferation of new state-level policies enacted over the past several years suggests that the next decade may see a dramatic and sustained increase in overall funding levels, and a fundamental re-drawing of the energy efficiency map. These new state energy efficiency policies reflect a variety of concerns, including the increasing cost and siting challenges of building new generation and transmission, fuel cost and supply risks, and the potential cost of future carbon regulations.

Within the past three years, for example, eleven states have adopted energy efficiency portfolio (or resource) standards (EEPS or EERS) that establish specific long-term savings targets that utilities are obligated to meet, and at least three other states are currently considering the same. A growing number of states have recently established laws requiring utilities to acquire all available cost-effective energy efficiency. Regulators in several Western states have also recently revised integrated resource planning (IRP) and demand-side management (DSM) planning rules to require more robust analysis of the resource potential and benefits of energy efficiency, which has resulted in increased savings targets for their energy efficiency portfolios (Hopper et al. 2008). Finally, regulators and utilities in many states are beginning to look more closely at regulatory incentive mechanisms to better align utility financial interests with improvements in customer energy efficiency.

We examined energy efficiency policies on the books or in the pipeline in all 50 states, along with recent IRPs and DSM plans, and developed low, medium and high projections of future energy efficiency spending and savings. Depending on how aggressively and effectively states implement these policies, we estimate that spending on ratepayer-funded energy efficiency could increase from $3.1 billion in 2008 to more than $12 billion (nominal dollars) per year by 2020 in our high case, a growth rate in spending of about 12% per year. Annual electricity savings nationally could triple from an estimated 0.3% of retail electricity sales in 2008 to 0.9% of retail electricity sales in 2020. In the low and medium scenarios, ratepayer funding for electric and gas energy efficiency in the U.S. would increase to $5.4 and $7.5 billion, respectively, by 2020.

What are the implications of such a scale-up of ratepayer-funded energy efficiency activity for national energy policy, such as a national EEPS or future carbon regulations? Can a ramp-up of this scale be achieved, and what practical constraints might slow these efforts?

This paper addresses these questions by first providing an overview of recent trends in state policies pertaining to ratepayer-funded energy efficiency programs in the U.S. The paper then presents our set of projections of future spending and savings from such programs, highlighting key themes. Projected energy savings are compared to what might be required under a future national EEPS (or broader clean energy standard that includes energy efficiency), in order to gauge the potential incremental impact of such policies. In addition, the carbon emission reductions associated with our projection of energy savings from ratepayer-funded programs is compared to the total emission reductions that might be required under the *American Clean Energy and Security Act of 2009* (aka, the Waxman-Markey bill), which was passed by the U.S. House of Representatives in June 2009 and would establish a cap on total greenhouse gas emission for many sectors of the U.S. economy. Last, the paper discusses some of the major obstacles and challenges that states and program administrators may face over the coming decade, as they seek to dramatically ramp-up ratepayer-funded energy efficiency program activity, as projected.

Table 1. 2008 Ratepayer-Funded EE Budgets

Rank	State	2008 Budget ($M)		
		Electric	Gas	Total
1	CA	831	183	1,014
2	NY	258	30	288
3	NJ	135	61	196
4	WA	160	18	179
5	MA	121	28	149
6	WI	76	64	140
7	MN	106	30	137
8	FL	109	15	124
9	CT	107	7	114
10	TX	106	no data	106
All Other States		592	94	686
U.S. Total		2,603	529	3,132

Source: CEE; excludes budget for load management programs.

2. Recent Developments and Trends in Ratepayer-Funding for Energy Efficiency Programs

Table 1 shows the 2008 approved electric and gas energy efficiency program budgets for the top- ten states (in terms of the absolute magnitude of the budget[3]), as well as the totals for the other 40 states and for the U.S. as a whole, based on data compiled by the Consortium for Energy Efficiency (CEE 2008).[4] Over 80% of the national total ($3.1 billion) came from electric ratepayers ($2.6 billion), while gas utilities were authorized to spend about $0.5 billion on energy efficiency programs. The 2008 budget total represents approximately a 20% increase over the $2.6 billion combined electric and gas energy efficiency budget in 2007. The top 10 states account for about 78% of the total U.S. energy efficiency budget, each of which planned to spend $100 million or more in 2008. California, the undisputed heavyweight in terms of the absolute magnitude of its spending on energy efficiency, represented one-third of the total U.S. energy efficiency program budget in 2008.

A proliferation of state-level energy efficiency policies adopted over the past several years suggests that the energy efficiency landscape in the U.S. may be on the verge of dramatic change, with an unprecedented expansion of ratepayer funding and a far-reaching geographical re-alignment of the energy efficiency marketplace. To understand these trends, it is helpful to classify states into three rough groupings: *Leaders*, *Up-and-Comers*, and *Uncommitted States*.

Leaders

Over the past 5-10 years, approximately 15 states have maintained relatively significant ratepayer-funding for energy efficiency programs (i.e., electric energy efficiency budgets equal to 1% or more of revenues from retail electricity sales, and annual electric savings equal to at least 0.5% of retail sales). Many of these leading states are listed in Table 1, although some have relatively small populations and thus have energy efficiency budgets that are small in absolute magnitude but large on a per-capita basis. In most leading states, the underlying policy support for energy efficiency program spending has been provided through integrated resource (or DSM- specific) planning and/or systems benefit funding mechanisms, and in some leading states, through statutory requirements that utilities acquire all (achievable) cost-effective energy efficiency savings opportunities.

Going forward, all leading states plan to maintain strong commitments to ratepayer-funded energy efficiency, and many have recently made (or are considering) commitments to substantially expand existing efforts. For example, New York established an aggressive energy efficiency portfolio standard (EEPS) requiring ratepayer-funded electric energy efficiency programs to achieve average annual savings of approximately 1.4% of retail sales from 2009- 2015, or more than double current levels. In pursuit of the new goals, the New York Public Service Commission authorized the state's investor-owned electric utilities to spend $330 million per year through 2011 on energy efficiency programs (NYPSC 2008). New Jersey also recently adopted aggressive new long-term energy savings goals that triple current savings levels, reaching approximately 2% of retail sales per year. The New Jersey Board of Public Utilities recently approved expanded budgets for the state's ratepayer-funded energy efficiency programs, rising to $325 million in 2012, in line with the long-term savings goals (NJBPU 2008). A new EEPS policy has also been proposed in Wisconsin, which would require electric savings of 2% of retail sales per year (approximately a quadrupling of current levels) by 2015. A number of other leading states (California, Connecticut, Massachusetts, Rhode Island, and Washington) have enacted statutes that require utilities to obtain all achievable cost-effective energy efficiency opportunities. To meet these mandates, state regulators in Massachusetts and

Connecticut are currently considering utility energy efficiency program plans or recommendations from appointed energy efficiency advisory boards that would double or triple 2008 spending levels by 2011, and would result in annual energy savings equivalent to 2-3% of retail energy sales (MEEAC 2009). In the Pacific Northwest, energy efficiency budgets are also likely to increase in the coming years, in order to meet the increasing energy efficiency savings targets identified in recent resource plans filed by major utilities in the region.

In California, the California Public Utilities Commission (CPUC) adopted a Long-Term Energy Efficiency Strategic Plan in September 2008, which provides a ten-year roadmap for a significant scaling up of the state's energy efficiency efforts and sets ambitious market transformation goals (CPUC 2008).[5] California's investor-owned utilities filed new energy efficiency program plans that propose roughly a $400 million (or 43%) increase by 2011 above the spending level authorized for 2008. The CPUC also adopted new long-term energy savings targets which likely will reduce the funding and savings trajectory somewhat for ratepayer-funded energy efficiency programs. The new savings targets represent about a 50% decline in ratepayer-funded efficiency savings by 2020 relative to 2008 levels. In adopting the lower savings targets, the CPUC relied to some extent on an updated statewide market potential study that forecasts lower remaining achievable energy efficiency savings potential.[6] In short, energy efficiency will remain a cornerstone of California energy policy for the foreseeable future, although the mix of strategies and the funding and savings levels may shift over time.

Up-and-Comers

There is a sizable contingent of states that, at least within recent history, have provided modest or little ratepayer-funding for energy efficiency, but have recently made significant commitments to ramping up energy efficiency programs.[7] Many of these up-and-coming states are located in the Midwest or Mid-Atlantic and have recently adopted aggressive EEPS policies, including: Illinois, Maryland, Michigan, New Mexico, Ohio, Pennsylvania, and Virginia (provisionally). The electric EEPS targets in most of these states rise to 1-2% of retail sales per year within the first 5-10 years of the standard, rivaling the annual savings levels currently being achieved in only a handful of leading states. North Carolina, another state that has, until now, been relatively inactive in the energy efficiency arena, enacted a renewables portfolio standard (RPS) under which energy efficiency can meet up to 40% of the total requirements of the state's investor owned utilities (IOUs) and an unlimited amount of the publicly owned utilities' requirements. Program development efforts in many of the aforementioned states are in their early stages, although the first set of energy efficiency plans have already been approved in Illinois and Maryland, with statewide budgets ramping up to $167 million and $91 million, respectively, over the first three years (ICC 2008a and 2008b; MDPSC 2008a, 2008b, 2008c, 2008d, 2008e).

A number of other up-and-comers have historically provided consistent, but moderate, levels of ratepayer-funding for energy efficiency and are in the process of dramatically ramping up program activity – in many cases, in response to recent policy mandates. Several of the up-and- comers in this group are located in the interior West. For example, Colorado adopted an EEPS with electricity savings targets for Xcel, the state's largest utility, reaching 1.3% of retail sales by 2020. The utility's recently-approved program plan puts it on track to meet its early-year EEPS targets, with a 2010 electric energy efficiency budget of $64 million (COPUC 2008). Nevada expanded its RPS in 2005 to allow energy efficiency to meet up to 25% of the total requirement, and program plans have been approved for the state's two IOUs that commit to achieving efficiency savings in excess of the maximum amount eligible under the RPS (NPC 2006, SPPC 2007). In Arizona, investor-owned utilities are now required to file DSM plans, and in its most- recent IRP, Arizona Public Service proposes to increase energy efficiency funding to $90 million in 2020, up from $23 million in 2008 (APS 2009).

Uncommitted States

Finally, in approximately 20 states, mostly located in the Southeast and parts of the Midwest, state regulators have not authorized utilities to fund energy efficiency programs, or have authorized only minimal funding levels, and have no immediate (public) plans to significantly step up their efforts. In some of these states, utilities do offer educational or general information programs related to energy efficiency, or offer audits to some customer classes or small pilot programs, but energy efficiency budgets are typically quite small (less than 0.1% of revenues), and programs are quite limited in scope. Going forward, some modest increase in ratepayer- funding for energy efficiency among uncommitted states is likely to occur in the face of escalating costs for new generation and the prospect of even higher costs for fossil fuel based generation as a result of future federal climate change legislation. Some utilities in these states have expressed increased interest in implementing energy efficiency programs, in many cases linking it to deployment of an advanced metering infrastructure (AMI) or approval to build new baseload generation.

Already, utilities and PUCs in a number of uncommitted states are considering moderate expansions to their energy efficiency efforts (in some cases contingent on developing an attractive business model for the utility), but have not yet made significant commitments.

3. Adding It All Up: Projected Ratepayer-Funded EE Spending

To gauge the potential impact of existing and impending state policies, legislative requirements, regulatory decisions and integrated resource or DSM plans (see Table 2), we developed a set of projections (low, medium, and high) of ratepayer spending on electric and natural gas energy efficiency programs through 2020. These projections do not account for the approximately $11 billion in federal funding for energy efficiency and related projects, made available through the American Recovery and Reinvestment Act (ARRA) to state and local governments, nor do the projections account for other "non-traditional" funding sources (e.g., revenues from emission auctions and from capacity market auctions). The potential significance of these additional funding sources is discussed briefly in Section V.

3.1. Approach

The projections for leading and up-and-coming states were developed based on varying state- or region-specific assumptions about how effectively and aggressively the energy efficiency policies currently in place (or under consideration) are implemented in the future (see Table 3).[8] For many of the leading states (with the exception of California, which is described separately in Table 3), the Low Case scenario assumes a continuation of current spending and savings levels – e.g., states achieve the spending/savings levels from the most-recently approved energy efficiency program plan or integrated resource plan, but do not exceed those levels in subsequent years. For the High Case, annual electric savings in leading states generally rises to at least 1.5%-2.5% of retail sales by 2020, depending on the region, with spending increasing to 3%-6% of revenues from retail electricity sales.

Most of the up-and-coming states have recently enacted EEPS policies. For those states (Colorado, Michigan, New Mexico, Pennsylvania) with EEPS savings targets that rise to a level consistent with what is currently being achieved in many leading states (i.e., roughly 1.0% of retail sales per year), we assume that the EEPS savings targets are achieved in both the Low and Medium cases, and that somewhat higher savings levels (1.2% to 1.5% of retail sales) are reached in the High Case scenario. For other up-and-coming states with EEPS policies (Ohio, Illinois, Maryland, and Virginia), we make more conservative assumptions, based on a number of different considerations.[9]

For the uncommitted states, we employ a standardized (and admittedly speculative) set of assumptions regarding the extent to which these states increase their support for ratepayer-funded energy efficiency programs in the future (see Table 3).

In each scenario, energy efficiency program *spending* is assumed to increase linearly to a target percentage of revenues by a specified year (0.3% of revenues by 2020 in the Low Case, 0.5% by 2012 in the Medium Case, and 0.8% by 2020 in the High Case). The High Case corresponds roughly to a scenario in which uncommitted states reach current national average spending levels by 2020 (and thus is well below what could be cost-effectively achieved in these states).

Somewhat greater spending levels are assumed for Florida compared to the other uncommitted states, as its current spending levels are moderately higher than the other states in this group. Also, for many leading and up- and-coming states, energy efficiency spending by publicly owned utilities (i.e., municipal utilities and rural electric cooperatives) was projected using the same scenario definitions as for the uncommitted states. This convention reflects the fact that publicly owned utilities are generally not subject to the same energy efficiency saving and spending policies as investor- owned utilities, and thus (in some states, at least) have historically provided less aggressive support for energy efficiency than their investor owned counterparts.

Table 2. Key Policy Drivers for Energy Efficiency Spending and Savings Projections

Policy Drivers	Applicable States*
Statutory requirement that utilities acquire all cost-effective energy efficiency	CA, CT, MA, RI, WA
EEPS	CA, CO, IL, MD, MN, MI, NJ (proposed), NM, NY, OH, PA, TX, VA (provisional), WI (proposed)
Energy efficiency eligibility under state RPS	HI, NC, NV
Recently-approved IRP plan	CO, ID, OR, MT, UT
Recently-approved DSM plan or multi-year budget	AZ, CT, CO, IA, MA, ME, NJ, RI (proposed), VT
System benefit charge	DC

* The Applicable States listed for each policy represents the states for which that policy informed development of spending and savings projections; it does not necessarily represent the totality of all states with that particular policy.

3.2. Results

Across the set of scenarios developed, ratepayer-funding for energy efficiency is projected to increase from \$3.1 billion in 2008 to \$4.1-\$8.8 billion in 2015 and to \$5.4-\$12.4 billion in 2020, depending on the scenario (see Table 4 and Figure 1). Under the Medium Case scenario, spending reaches \$6.0 billion in 2015 and \$7.5 billion in 2020. As a percentage of total revenue from electricity and natural gas retail sales to end-use customers, spending in 2020 is projected to reach 0.8% of total revenues in the Low Case, 1.1% in the Medium Case, and 1.8% in the High Case – compared to 0.6% in 2008. Under all three scenarios, funding for electric energy efficiency is expected to represent approximately 90% of the total for electric and natural gas energy efficiency, which is roughly consistent with the historical funding distribution and is indicative of the greater emphasis of recent state energy efficiency policies on the electric sector.

Table 3. Scenario Assumptions Underlying Electric EE Spending and Savings Projections

Group of States	Scenario	Representative Assumptions (specific assumptions vary by state)
Leaders (non-CA)	*Low*	States generally meet existing EEPS targets or savings levels identified in approved IRPs or maintain spending at existing levels, based on current SBC or approved DSM plan
	Medium	Highly state-specific: an approximate doubling of current spending for many New England states by 2020; no change from low case for some Northwestern states; NY meets its EEPS
	High	States generally reach benchmark 2020 savings level for region (2.0-2.5% of retail sales for most New England states, 1.5% for Northwest states); NY spending reaches 3.5% of revenues by 2020; NJ and WI adopt and meet proposed EEPS targets
California	*Low*	IOU annual savings equal to current long-term savings goals through 2020; POUs' annual savings equal 75% of the average annual savings identified in their 10-yr. EE plans
	Medium	IOU savings same as low case; POUs attain 100% of savings levels identified in their 10-yr plans
	High	IOUs maintain savings at 2008 goal level through 2020; POU savings are same as in medium case
Up-and-Comers	*Low*	Similar to low case assumptions for Leading states, but EEPS targets are not achieved in IL or OH
	Medium	No change from low case for many states; 2020 savings increase to approx. 0.7%-1.0% of retail sales in several others
	High	Savings in most states reach 1.0-1.5% of retail sales by 2020; IL fully meets its EEPS targets; TX reaches annual savings of 1.0% of retail sales by 2020
Uncommitted States	*Low*	Spending increases linearly from 0.1% of revenues in 2008 to 0.3% in 2020; FL spending remains flat at 0.4% of revenues from 2008-2020
	Medium	Spending increases linearly from 0.1% of revenues (0.4% in FL) in 2008 to 0.5% in 2012, and remains flat thereafter
	High	Spending increases linearly from 0.1% of revenues (0.4% in FL) in 2008 to 0.8% in 2020

Table 4. Range in Ratepayer-Funding Projections for 2015 and 2020

	Spending ($billion, nominal)			Spending (% of revenues)		
	2008*	2015	2020	2008*	2015	2020
Electric	2.6	3.7 - 7.8	4.9 - 10.9	0.7%	1.0% - 2.0%	1.0% - 2.2%
Gas	0.5	0.4 - 1.0	0.5 - 1.6	0.3%	0.3% - 0.7%	0.2% - 0.8%
Total	3.1	4.1 - 8.8	5.4 - 12.4	0.6%	0.8% - 1.6%	0.8% - 1.8%

* 2008 data are approved budgets, rather than actual spending; source for 2008 dollar values: CEE (2008).

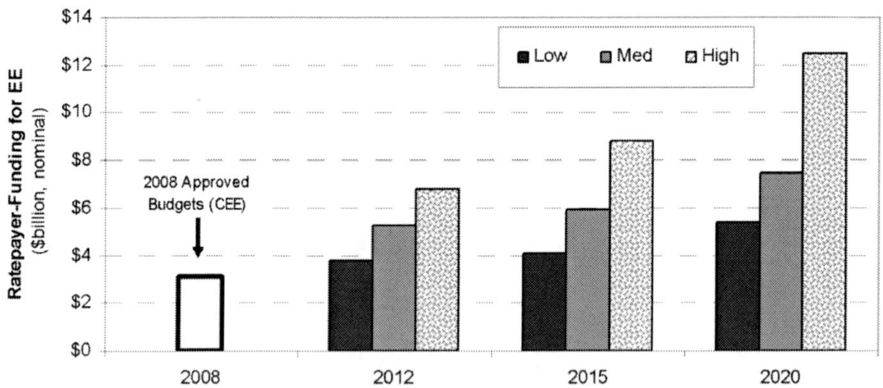

Figure 1. Projected Ratepayer-Funding for Electric and Natural Gas Energy Efficiency Programs in the U.S.

Broad adoption of new state-level energy efficiency policies also raises the prospect for significant geographical re-alignment of energy efficiency markets. Table 5 lists the 10 states with the highest projected levels of ratepayer-funding in 2020 under the Low and High Case scenarios, compared to the top 10 states in 2008, along with aggregate spending in the other 40 states. Several specific trends are worth highlighting. First, as new state energy efficiency portfolio standards are implemented, a number of states with large populations that, heretofore, have been relatively minor players on the national energy efficiency stage (e.g. PA, IL, OH, MI) assume much more prominent roles by 2020. Second, funding levels become markedly more evenly distributed, both within the top tier of states. For example, all of the top-10 states (in terms of the absolute magnitude of spending) reach projected 2020 funding levels in excess of $200M/yr in the Low Case and $400M/yr in the High Case, and the gap between California and other top-10 states narrows considerably compared to 2008. Funding also becomes considerably more evenly distributed at the broader, national level, with more than 40% of projected funding in 2020 occurring outside of the top-10 states, compared to 22% today.

Recently adopted energy efficiency policies in many states will require quite dramatic increases in spending over the next decade, which, for up-and-coming states, will have to be built from a quite modest base. Table 6 lists the 10 states with the largest projected funding increases from 2008 to 2020. In the Low Case, spending increases by $100M-$300M in 10 states over the decade. Only three of these states (NY, MA, and NJ) had significant energy efficiency budgets in 2008; the others are all up-and-coming states with recently enacted EEPS policies (or, in North Carolina, an RPS policy in which energy efficiency is a qualifying resource). In the High Case, more than 10 states register spending increases in excess of $300M by 2020, with three large states (NY, TX, and IL) seeing an increase of greater than $750M. In a later section of this paper, we discuss some of the issues that states may confront as they seek to dramatically ramp- up energy efficiency program activity.

Table 5. Top-10 Energy Efficiency Markets in 2020, Ranked by Annual Budget Projections

Rank	2008 Budget ($M, nominal)		2020 Spending Projections Low ($M, nominal)		2020 Spending Projections High ($M, nominal)	
1	CA	1,014	CA	492	CA	1,312
2	NY	288	NY	477	NY	1,094
3	NJ	196	OH	350	TX	882
4	WA	179	NJ	344	IL	805
5	MA	149	MA	297	MA	630
6	WI	140	NC	273	OH	595
7	MN	137	PA	247	WI	575
8	FL	124	MI	238	NJ	504
9	CT	114	IL	230	PA	467
10	TX	106	WA	202	MN	413
Top-10 ($M)	2,447		3,150		7,276	
% of U.S.	78%		59%		58%	
Other States ($M)	686		2,211		5,168	
% of U.S.	22%		41%		42%	

Table 6. States with Largest Projected Funding Increase (2008-2020)

Rank	Low Case ($M, nominal)			High Case ($M, nominal)		
	State	2008 Budget	2008-2020 Increase	State	2008 Budget	2008-2020 Increase
1	OH	58	292	NY	288	806
2	NC	0	273	TX	106	775
3	PA	0	247	IL	41	764
4	MI	20	218	OH	58	537
5	NY	288	189	MA	149	481
6	IL	41	189	PA	0	467
7	MA	149	148	WI	140	435
8	NJ	196	148	MD	6	348
9	MD	6	143	NC	0	324
10	CO	26	141	MI	20	313

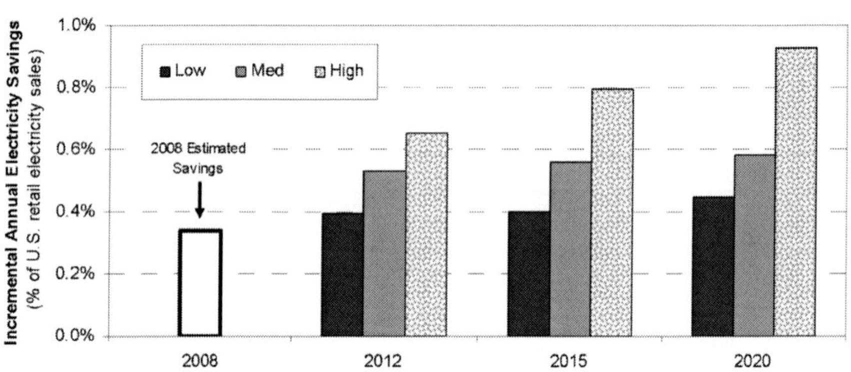

Figure 2. Projected Incremental Annual Electric Energy Efficiency Savings from Ratepayer-Funded Programs in the U.S.

4. PROJECTED ELECTRICITY SAVINGS IN PERSPECTIVE

In 2008, electric energy efficiency programs are estimated to have resulted in incremental annual savings of approximately 13 billion kWh, equivalent to 0.34% of total retail electricity sales nationally in that year.[10] Several leading states have achieved incremental annual savings greater than 1.0% of retail energy sales, and VT reported annual energy savings in 2008 equivalent to 2.5% of retail energy sales (Efficiency Vermont 2009). Projected increases in ratepayer spending on energy efficiency programs over the next decade are expected to result in significant increases in energy savings. Figure 2 shows our projections of incremental annual savings as a percentage of retail electricity sales, for the three scenarios. From 2010-2020, average annual savings range from 0.41% of retail sales in the Low Case to 0.76% in the High Case. In comparison, EIA's April 2009 reference case forecast (which does not *explicitly* account for ratepayer-funded energy efficiency) projects that retail electricity sales will grow at an average rate of 1.1% per year over this period. *Cumulatively*, electricity savings achieved from ratepayer-funded programs implemented over the 2010-2020 period are projected to reduce U.S. electricity consumption in 2020 by about 4.7% (relative to EIA's reference case forecast of 2020 retail electricity sales) in the Low Case, and by about 8.6% in the High Case. Approximately 90% of all savings is projected to occur within leading and up-and-coming states that have already made a significant commitment to supporting ratepayer-funded energy efficiency.

4.1. Incremental Impact of a National EEPS or Clean Energy Standard

Several national EEPS proposals have been introduced in the U.S. Congress in 2009, and several proposed national clean energy standards – including the *American Clean Energy and Security Act of 2009*, which recently passed the House of Representatives – would allow energy efficiency to be used to meet a portion of the overall target in each year. Our projections of state-level energy efficiency program savings do not explicitly account for the effects of future national energy efficiency legislation and can therefore serve as a benchmark for assessing the potential *incremental* impact of such policies.

Table 7. Projected Incremental Impact of a National EEPS on Ratepayer-Funded Energy Efficiency Program Savings

National EEPS Saving Target: Cumulative Savings in 2020 as a Percent of Retail Sales	Percentage Increase in EE Program Savings Relative to No National EEPS
5%	0% - 12%
10%	8% - 37%
15%	18% - 68%

We estimate the incremental increase in savings from ratepayer-funded electric energy efficiency programs that could occur under several "generic" national EEPS (or clean energy standards) policies, which are modeled on recent federal legislative proposals. Under the generic policies, all retail electricity distributers with greater than 1 million MWh sales per year, on average, would be required to meet specified cumulative electric savings targets by 2020. We consider three potential cumulative savings targets, equal to 5%, 10%, and 15% of retail electricity sales in 2020. Only savings from measures implemented in 2010 or later would count towards the targets. Compliance could be achieved through utility energy efficiency programs, new building energy codes, new appliance efficiency standards, new combined heat and power, and distribution system upgrades. The policy would not create a national market for tradable energy efficiency certificates; however, it would allow obligated retail electricity distributers to meet a portion of their EEPS obligations by purchasing electricity savings from measures implemented in the same state, via bilateral contracts with other distribution utilities, state agencies, or other energy efficiency program providers.

To determine the incremental increase in savings from ratepayer-funded electric energy efficiency programs that would be required under these generic EEPS proposals, we assume that *half* of the total energy savings target in each state would be met through ratepayer-funded programs, and the remaining half would be met through a combination of codes and standards, CHP, distribution system upgrades, and other eligible measures. We further assume that bilateral contracting for electricity savings from measures implemented within each state occurs freely without constraint. Given these assumptions, we estimate that a 5% national EEPS would require a 0-12% increase in savings from ratepayer-funded energy efficiency programs (where the lower percentage number represents the increase relative to our High Case savings projection, and the higher percentage number represents the increase relative to our Low Case savings projection). Similarly, we estimate that a 10% national EEPS would require an 8-37% increase, and a 15% national EEPS would require an 18-68% increase (see Table 7). If states were to rely more heavily on utility programs than we assume, or if intra-state bilateral contracting of energy efficiency savings for compliance with the national EEPS were constrained, then larger percentage increases would be needed. Moreover, the percentage savings increases presented here represent the aggregate national increase. Naturally, the additional savings required in individual states may be substantially higher or lower than the national average, depending on what portion of the state's retail electricity sales is exempt from the EEPS, and on the amount of savings that would occur in the absence of a national EEPS.

Although our analysis is based on a single generic national EEPS policy, it does reveal that a relatively aggressive national EEPS (e.g., a cumulative savings target on the order of 15% of retail sales by 2020) could require substantial increases in energy savings, even in the face of the spate of aggressive state policies that have recently been adopted. The impact, however, would likely fall primarily on about 20 or so states that have not yet made significant commitments to supporting ratepayer-funded energy efficiency. In addition, the incremental impact of a national EEPS depends heavily on its design – e.g., overall target levels, the set of eligible measures, and allowances for bilateral contracting and/or trading of energy savings certificates. In particular, allowing unrestricted trade of energy savings certificates, without a sufficiently aggressive target, could substantially degrade the incremental impact of a national EEPS.[11]

4.2. Contribution to Greenhouse Gas Emission Reduction Targets

Understanding the potential range of energy savings resulting from ratepayer-funded efficiency programs over the next decade is also important from the perspective of gauging their contribution to future carbon emission reduction targets that Congress may adopt. For example, *The American Clean Energy and Security Act of 2009*, which recently passed the U.S. House of Representatives, would place a cap on greenhouse gas emissions from certain sectors of the U.S. economy.[12] Based on an analysis of the bill by the Environmental Protection Agency (EPA), the 2020 cap would require a reduction of 912 million metric tons of carbon dioxide equivalent (mmtCO2e) in covered sectors, relative to projected 2020 emissions under the reference case scenario.[13] EPA's reference case is based on the Energy Information Administration (EIA)'s April 2009 update to its 2009 Annual Energy Outlook reference case forecast. Although EIA's forecast does not explicitly account for electricity savings from ratepayer-funded energy efficiency programs implemented over the forecast period, it is possible that some fraction of the savings from such programs may be implicitly captured within the forecast. Lacking specific knowledge, if we assume that the reference case includes a continuation of ratepayer-funded energy efficiency savings at 50% of the level achieved in 2008, then our projections of electricity savings from ratepayer-funded programs would yield emission reductions in the range of 42-164 mmtCO2e by 2020, relative to the reference case.[14] Though not a negligible impact, it is clear that even the fairly aggressive increases in energy efficiency spending and savings that we project over the next decade would contribute a relatively moderate percentage to the overall emission reductions that future federal climate regulations might potentially require (e.g., 912 mmtCO2e by 2020 under *The American Clean Energy and Security Act of 2009*). That said, ratepayer-funded energy efficiency programs represent just one strategy for acquiring energy efficiency savings; accounting for the effects of new building codes, appliance efficiency standards, tax incentives and other federal programs would increase the total contribution of energy efficiency towards any future greenhouse gas emission reduction goals.

5. Challenges Ahead, Bumps in the Road?

Is a doubling or tripling of ratepayer-funding for energy efficiency over the next decade feasible? What are the fundamental challenges that confront states' efforts to achieve increasing levels of savings from ratepayer-funded programs? Below, we highlight several issues that we believe may dominate the landscape on the road ahead.

The Economic Downturn

At least over the near-term, economic conditions in the U.S. may complicate and constrain efforts to scale-up energy efficiency spending and savings, for several reasons. First, the success of ratepayer-funded energy efficiency programs typically requires customers to pay a portion of the capital outlay for energy efficiency measures. As households and businesses struggle to manage their day-to-day expenses, many may be reluctant make new investments, even those with short payback periods. Second, a slowing or decline in overall economic activity will likely reduce the rate of stock turnover and new housing starts, thereby reducing the amount of energy savings that could be captured through ratepayer-funded programs targeting these market opportunities. Finally, economic hard-times may complicate the political feasibility of increasing ratepayer-spending for energy efficiency, particularly if it results in short-term rate impacts.[15]

Aversion to Short-Term Rate Impacts from Large-Scale Energy Efficiency Implementation

In most states, utilities typically expense program costs for energy efficiency as they are incurred, hence cost recovery is relatively front-loaded compared to cost recovery for most supply-side resource alternatives. As a result, the rate impacts from energy efficiency tend to occur sooner (even if the rate impacts are less over the long-term, and even if average utility bills are reduced compared to supply-side alternatives). Thus, the short-term rate impacts associated with attaining very aggressive levels of savings could pose a political challenge for state regulators, particularly in those states that have seen significant rate hikes in recent years or whose rates are well-above national averages. This issue has surfaced in several states that have recently passed an EERS (e.g. IL, PA) as they have established cost caps or limits on rate impacts. To mitigate concerns about short-term rate impacts, we expect that an increasing number of state PUCs and utilities will explore cost recovery mechanisms that allow better alignment of program costs and system benefits (e.g. amortization of expenses) as well as decoupling and shareholder incentive mechanisms.

Coordination with State/Federal Energy Efficiency Programs

In most states and market segments, ratepayer-funded energy efficiency programs administered by utilities or third parties have been implemented at a much larger scale than state or Federal energy efficiency programs (except for low-income weatherization). The passage of *The American Recovery and Reinvestment Act* (ARRA) is clearly a "game-changer" in terms of the level of support and funding provided by Federal taxpayers for energy efficiency over the next three years. Funding for the federal low-income weatherization program ($5 billion) and State Energy Program (SEP) formula grants ($3.1 billion targeting energy efficiency retrofits and renewable energy in buildings and industrial facilities) will increase ten-fold from current levels, and the Energy Efficiency and Conservation Block Grant Program will provide an additional $3.2 billion in grants to local governments for various energy efficiency and renewable energy activities. The intent of the SEP formula grants, specifically, is for states to commit their grant funding to expand existing programs, including ratepayer-funded programs, or to create new programs – not to supplant or replace existing funding (U.S. DOE 2009).

Thus, state and local energy efficiency programs that are implemented with ARRA funds will require a massive ramp- up in capability to deliver energy efficiency services.

Because we are primarily interested in forecasting longer-term trends in ratepayer-funded energy efficiency programs, our analysis does not explicitly account for the potential short-term impacts of ARRA-funded state and local EE programs on existing and emerging ratepayer-funded EE programs. However, the infusion of funding for energy efficiency provided by ARRA does present several near-term implications for ratepayer-funded programs. First and foremost, a much higher degree of coordination will be required among energy efficiency program administrators in each state (e.g. utilities, state energy office, local governments) in order to ensure consistency in program offerings, obtain support from trade allies, and minimize confusion among customers and program delivery contractors. Because customers will have access to multiple funding sources to buy-down the cost of their energy efficiency projects (e.g. ratepayer and ARRA funds), state PUCs will have to develop guidelines regarding whether program administrators will be able to take full or partial credit for savings funded by different programs.[16] In addition, if state PUCs perceive that end users and the service delivery infrastructure will not be able to absorb the massive increase in funds for energy efficiency because of the ARRA, then some states may delay or slow down their internal plans to ramp-up spending on ratepayer-funded energy efficiency programs until after ARRA funds are committed or spent (2012).

Developing Innovative Program Designs to Reach Deeper and Broader Savings

Given the historic experience of leading states, we are confident that the 20 or so uncommitted states can achieve the energy efficiency savings levels in our High Case scenario (i.e., annual savings of approximately 0.3% of electric retail sales by 2020), because many utilities across a diverse set of circumstances are currently implementing programs that produce significantly greater levels of savings on a sustained basis. However, a number of leading and up-and-coming states have established aggressive energy efficiency savings goals for future years that are well beyond current experience and practice in most leading states (e.g., annual electric savings on the order of 2% or more of retail sales). Based on recent market potential studies in several of these states, savings at this level are achievable and cost-effective compared to supply-side alternatives. However, the challenge for program administrators will be to design and implement programs that can achieve both deeper savings, on average, at customer facilities and have a broader reach in terms of market penetration, over a sustained period of time. Service providers will have to achieve savings levels of 25-40% of existing usage at customer facilities compared to current practice in ratepayer-funded programs, which is typically in the 5-20% range. Achieving higher market penetration rates will require programs to target and reach traditionally under-served markets (e.g. small commercial, multi-family, rental housing, non-owner occupied commercial facilities) in far greater numbers than current practice (MEEAC 2009). We are also likely to see increased attention to integrated delivery of electric and gas efficiency programs as well as coordinated delivery of energy efficiency and on-site renewables and combined heat and power, in order to reduce transaction costs and provide customers with tailored, customized service offerings.

Greater Reliance on Aggressive Appliance and Lighting Efficiency Standards

Ratepayer-funded (or taxpayer-funded) energy efficiency programs, federal and state appliance efficiency standards, and building codes represent three major strategies to capture energy efficiency savings potential. These three strategies can complement each other but also are partial substitutes, in the sense that adoption of aggressive minimum efficiency standards for a broad range of consumer products raises the current-practice baseline and may therefore reduce the remaining achievable potential that can be captured by voluntary, ratepayer-funded energy efficiency programs.

For example, in the 2007 energy bill, new standards were adopted for 10 products, including general service lamps, which are required to use about 25-30% less energy than current incandescent bulbs by 2012-14 and 60% less by 2020.[17] When fully phased in, the new standard for general service lamps, combined with increasing market penetration of compact fluorescents lighting (CFLs), may mean that CFLs applications that are eligible measures in many ratepayer- funded programs will have become standard practice. Residential lighting measures often account for 20-40% of total savings in a utility's energy efficiency portfolio and are among the most cost-effective measures. Thus, in the future, program administrators will likely be required to support new measures (e.g., LED lighting) and develop new strategies to capture cost-effective savings opportunities, because some of the measures implemented currently in programs (e.g. standard CFLs) may be covered by appliance efficiency standards.

Our analysis of future spending and savings scenarios for ratepayer-funded EE programs does not explicitly account for the possibility that federal and/or state governments could decide to significantly revise the balance among existing EE policies and decide to rely more heavily on building energy codes and efficiency standards to capture the available savings potential. We recognize that this is an important issue, though, and note that the California PUC, in adopting new savings goals for ratepayer-funded energy efficiency programs for the 2009-2020 period, explicitly accounted for the impacts of new federal standards and state codes and standards. The CPUC correspondingly reduced the remaining market potential for ratepayer-funded energy efficiency programs. It is possible that energy efficiency goals in other states have not fully accounted for the impact of federal appliance standards, if these goals were based on market potential studies conducted prior to enactment of those standards. At the same time, new efficiency opportunities may emerge over the coming decade that will replenish the market potential reservoir.

Developing the Institutional Framework for Energy Efficiency in Up-and-Coming and Uncommitted States

Energy efficiency resources have some distinctive characteristics that require state PUCs to establish an institutional framework for effective regulatory oversight of ratepayer-funded energy efficiency programs. These distinctive elements include: the need for measurement and verification of savings from energy efficiency programs; the fact that energy efficiency programs depend on customer acceptance and adoption; and the fact that, under traditional regulation, the utility's financial interests are not well-aligned with societal interest in pursuing cost-effective efficiency. Many leading states have successfully grappled with these institutional and regulatory policy issues, and a variety of approaches have proven to be effective. Based on this experience, state PUCs must provide leadership in defining energy efficiency policy objectives, establish roles and responsibilities for program administrators, and be willing to devote sufficient staff (or technical consultant) resources so that they can effectively oversee acquisition of large-scale energy efficiency resources.

Workforce Needs and Human Infrastructure

The projected growth in ratepayer- and taxpayer-funded energy efficiency will require significant near- and long-term expansions of the energy efficiency services sector (EESS) workforce. Trained personnel are and will be needed to design, implement, and manage energy efficiency programs, and to design, construct, install and maintain energy efficiency building systems. There is growing concern among policymakers and program administrators about a shortage of well-trained personnel in the EESS. Goldman et al. (2009) attempted to estimate the size and structure of the current EESS workforce, forecast the number of people that will be needed in 2015 and 2020 under various scenarios, and assess shortages and needs in the current workforce in light of expectations for growth (Goldman et al. 2009).

Drawing upon over 300 interviews, survey respondents reported that:

- management positions requiring at least 10 years experience and positions requiring engineering experience with high-efficiency technologies are the most difficult positions to fill (taking up to 12-15 months to fill);
- re-training of the existing construction and building industry workforce to provide energy efficiency services will be critical to success (and that this workforce is "under-employed" given the economic downturn);
- aging workforce is viewed as a problem among building and construction industry association members (40-45% of workforce is over 50) but not in other EESS workforce segments;
- there is a shortage of engineers with efficiency knowledge or experience and energy efficiency services employers are thus willing to hire any engineer with technical aptitude, communication skills and some engineering experience;
- there are not enough certificates or degrees being awarded in energy efficiency related training and education to meet the growing need, although the training/education capability is expanding.

In our view, the combined impact of ratepayer-funded programs and energy efficiency programs that will be funded by the ARRA means that the shortage of well-trained personnel is primarily a near-term issue (e.g. 1-3 years). By 2013, if energy efficiency programs that are funded by the ARRA wind down, there will be a much larger workforce base in the EESS, which should be able to meet workforce demands that would occur under our High Case projection of ratepayer- funded energy efficiency program spending.

CONCLUSION

A proliferation of new state-level energy efficiency policies enacted over the past several years suggests that ratepayer-funded energy efficiency program savings and spending will continue to grow at a brisk pace over the next decade, and that the geographical structure of the energy efficiency market will undergo a fairly dramatic re-alignment. Based on a bottom-up analysis of state-level energy efficiency policies, we project that ratepayer-funding for electric and gas energy efficiency programs will rise from $3.1 billion in 2008 to between $5.4 billion (Low Case) and $12.4 billion (High Case) by 2020, with a Medium Case estimate of $7.5 billion. Along with the increase in aggregate spending, we project a significant broadening of the national energy efficiency market. A large portion of the overall projected increase will be centered in relatively populous states that, until recently, have been relatively minor players on the national energy efficiency stage (e.g., IL, MI, NC, OH, and PA). Thus, for example, we project in our Medium Case that 15 states will have annual energy efficiency budgets greater than $200 million in 2020, compared to just 2 states in 2008.

Our aggregate U.S. spending projections equate to average annual growth rates of 4.6% (Low Case), 7.5% (Medium Case), and 12.2% (High Case). In comparison, spending grew by 18% per year, on average, from 2006-2008. Moreover, the projections do not account for recent federal stimulus funding (ARRA), the potential impact of future national energy efficiency legislation (e.g., a national EEPS or a national RPS with an energy efficiency component) or future federal climate change legislation. Thus, our Medium Case is arguably a somewhat conservative estimate of the "most likely" spending trajectory, and our High Case projection is a more probable outcome than the Low Case.

That said, states are likely to face a number of key challenges as they seek to ramp up energy efficiency program activity. Some of these challenges are primarily near-term issues – e.g., the economic downturn, shortages within the energy efficiency workforce, and the (sometimes painstaking) process of developing the requisite institutional infrastructure and regulatory oversight mechanisms in states that are in the early phases of implementing ratepayer-funded efficiency programs. Other issues are ongoing or longer-term, such as the need to develop new and innovative program designs to achieve deeper and broader energy savings, in order to reach state energy savings goals that significantly exceed what is currently being achieved. This challenge will become increasingly pertinent as new appliance efficiency standards and building codes deplete some of the traditional low-hanging fruit for voluntary energy efficiency programs. While significant, we believe that these challenges are surmountable.

TECHNICAL APPENDIX: METHODOLOGY AND ASSUMPTIONS USED TO DEVELOP SPENDING AND SAVINGS PROJECTIONS

The paper presents the projections, developed by Lawrence Berkeley National Laboratory, of ratepayer-funded energy efficiency program spending (electric and natural gas) and savings (electric only) through 2020. This technical appendix describes the methods and assumptions used to develop these projections.

A1. Electric Energy Efficiency Spending & Savings Projections

Low, medium, and high projections of future electric energy efficiency program savings and spending were developed on a state-by-state basis. Although many of the specific assumptions and the approach to defining scenarios varied by state, the basic methodology used in all states consisted of several common components, including:

- Developing projections of retail electricity sales and revenues from retail electricity sales
- Defining low, medium, and high scenarios of future ratepayer-funded energy efficiency program savings and spending for the electricity sector
- Defining the amount of spending required to achieve different levels of savings

Each of these elements is described further below.

A1.1. Retail Sales and Revenue Projections

Projections of annual retail electricity sales and revenue from retail electricity sales were used to develop energy efficiency program spending and savings projections, and were also used to develop metrics that allow for comparison of spending and savings levels across states of differing sizes (e.g., savings as a percent of retail sales and spending as a percent of revenues).

Baseline Retail Sales and Revenue Projections

An initial set of baseline retail sales and retail price projections for each state was developed by applying annual growth rate projections from the Energy Information Administration (EIA)'s April 2009 update to its Annual Energy Outlook 2009 (AEO2009) reference case forecast to actual 2007 retail sales and price data for each state, as reported on EIA's Form-860. The electricity retail sales and retail price projections in AEO2009 are specified at the Electricity Market Module (EMM) level, the regions used in EIA's National Energy Modeling System (NEMS). Thus, the EMM-level growth rates were applied to each state in the respective region. Table A-1 summarizes the annual average growth rates (2008-2020) of retail electricity sales and retail electricity prices in each EMM region, from the April 2009 AEO2009 reference case forecast. Revenue projections were calculated by multiplying projected retail electricity prices by projected retail electricity sales, and were converted to nominal dollars using the AEO2009 reference case forecast of the GDP chain-type price index.

State-Specific Adjustments for Each Scenario

Future retail sales and revenues in each state will depend, in part, on the amount of savings achieved from future ratepayer-funded energy efficiency programs. In order to maintain internal consistency, we adjusted the retail sales and revenue projections for each scenario in each state, to reflect the energy efficiency savings assumed for the given scenario. The adjustments consisted of increasing or decreasing the baseline sales and revenue in each year, to account for the cumulative difference between the savings assumed for the scenario and the savings assumed to be implicit in the AEO2009 forecast.

The NEMS model does not explicitly account for ratepayer-funded energy efficiency programs; however, it does model future energy efficiency improvements at the end-use level, which may be partly attributable to future ratepayer-funded energy efficiency programs. Lacking better information, we assumed that the baseline retail sales projections, derived from the AEO2009 forecasted growth rates, implicitly account for a continuation of ratepayer-funded energy efficiency programs with savings equal to 50% of historical levels, as summarized in Table 2. Historical savings at the level of each EMM region were calculated from data on actual savings achieved in each state from ratepayer-funded electric efficiency programs implemented in 2006, as compiled by ACEEE (Eldridge et al. 2008). To provide an example: if we project, under one scenario, that future savings in a given state will be equal to 0.3% of retail sales in each year, and the energy efficiency savings assumed to be implicit in the baseline retail sales forecast is 0.1%, then we would reduce the forecast in each year to account for the cumulative effect of the additional 0.2% of retail sales saved each year (i.e., reduce the retail sales projection by 0.2% in year 1, by 0.4% in year 2, and so on).

Table A-1. AEO2009 Projected Growth Rates in Retail Electricity Sales and Prices

EMM Region	States	AEO2009 Average Annual Growth Rate (2009-2020)	
		Retail Electricity Sales	Retail Electricity Price (real)
East Central Area Reliability Coordination Agreement	IN, KY, MI, OH, WV	0.6%	-0.2%
Electric Reliability Council of Texas	TX	0.9%	-0.3%
Mid-Atlantic Area Council	DC, DE, ME, NJ, PA	0.7%	-0.4%
Mid-America Interconnected Network	IA, IL, MO, WI	0.7%	0.1%
Mid Continent Area Power Pool	MN, ND, NE, SD	0.8%	0.7%
Northeast Power Coordinating Council/New York	NY	0.5%	-0.1%
Northeast Power Coordinating Council/New England	CT, MA, ME, NH, RI, VT	0.7%	-0.8%
Southeastern Electric Reliability Council/Florida	FL	1.3%	0.6%
Southeastern Electric Reliability Council/ Excluding Florida	AL, AR, GA, LA, MS, NC, SC, NT, VA	0.9%	0.0%
Southwest Power Pool	KS, OK	0.9%	0.1%
Northwest Power Pool Area	ID, MT, OR, UT, WA, WY	1.0%	0.1%
Mountain Power Area, Arizona, New Mexico, and Southern Nevada	AZ, CO, NM, NV	1.5%	0.2%
California	CA	0.9%	-0.9%

Table A-2. Assumed Baseline Energy Efficiency Savings in AEO2009 Forecast

EMM Region	Baseline Ratepayer-Funded Energy Efficiency Program Annual Savings (% of Retail Sales)
East Central Area Reliability Coordination Agreement	0.0%
Electric Reliability Council of Texas	0.1%
Mid-Atlantic Area Council	0.0%
Mid-America Interconnected Network	0.1%
Mid Continent Area Power Pool	0.2%
Northeast Power Coordinating Council/New York	0.3%
Northeast Power Coordinating Council/New England	0.4%
Southeastern Electric Reliability Council/Florida	0.1%
Southeastern Electric Reliability Council/Excluding Florida	0.0%
Southwest Power Pool	0.0%
Northwest Power Pool Area	0.3%
Mountain Power Area, Arizona, New Mexico, and Southern Nevada	0.1%
California	0.4%

A1.2. Scenario Definitions

In order to simplify the scenario-development process, the 50 states and Washington D.C. were segmented into two groups (see Table 3). States in Tier I are those that either have a strong history of supporting ratepayer-funded energy efficiency programs (*Leaders*) or have recently enacted policies that will likely provide strong support going forward (*Up-and-Comers*). As described further below, the scenarios for these states were developed based on largely state- specific assumptions reflecting policies or plans currently in place or under consideration. For the remaining states (*Uncommitted States*), in Tier II, spending and savings projections were developed using a relatively simple, standardized approach. Spending/savings projections for municipal utilities and cooperatives in many Tier I states were developed using the same approach as used for Tier II states.

Table A-3. Segmentation of States into Tier I and Tier II for Electric Efficiency Projections

Tier I	AZ, CA, CO, CT, DC, DE, HI, IA, ID, IL, MA, MD, ME, MI, MN, MT, NC, NH, NJ, NM, NV, NY, OH, OR, PA, RI, TX, UT, VA, VT, WA, WI
Tier II	AK, AL, AR, FL, GA, IN, KS, KY, LA, MO, MS, ND, NE, OK, SC, SD, TN, WV, WY

Table A-4. Key Policy Drivers for Tier I-State Spending and Savings Scenarios

Policy Drivers	Applicable States*
Statutory requirement that utilities acquire all cost-effective energy efficiency	CA, CT, MA, RI, WA
EEPS	CA, CO, IL, MD, MN, MI, NJ (proposed), NM, NY, OH, PA, TX, VA (provisional), WI (proposed)
Energy efficiency eligibility under state RPS	HI, NC, NV
Recently-approved IRP plan	CO, ID, OR, MT, UT
Recently-approved DSM plan or multi-year budget	AZ, CT, CO, IA, MA, ME, NJ, RI (proposed), VT
System benefit charge	DC

* The Applicable States listed for each policy represents the states for which that policy informed development of spending and savings projections; it does not necessarily represent the totality of all states with that particular policy.

Tier I Scenario Definitions

Low, medium, and high case scenarios were developed individually for each Tier I state. The starting point in defining the scenarios was the set of energy efficiency policies currently in place (or under consideration). Table 4 identifies the set of policies considered, and the states for which those policies were used as an input to one or more scenario. Table 5 describes the specific assumptions underlying each scenario in each Tier I state, and indicates more specifically how the policies listed in Table 4 were applied across the scenarios.

Table A-5. Scenario Descriptions for Each Tier I State

State	Case	Scenario Description*
AZ	Low	IOUs maintain savings at 0.8% of retail sales (equal to the 2010 net savings level in APS' approved 2008-2010 DSM plan).
	Medium	Same as Low Case
	High	IOUs ramp up savings from 0.8% of retail sales in 2010 to 1.5% in 2020.
CA	Low	IOU savings equal to CPUC savings goals through 2020 (consisting of the 2004-2013 goals established in 2004, and the interim 2012-2020 goals established in 2008, which supersede the old goals for the years 2011-2012). POU annual savings are equal to 75% of the average annual savings from their 2007-2016 EE plans.
	Medium	Similar to Low Case, except POUs are assumed to attain 100% of the savings in their EE plans.
	High	Similar to Medium Case, except IOU savings from 2012-2020 remain at average annual savings of original CPUC IOU goals for 2007-2013, equal to 1.1% of retail sales.
CO	Low	IOUs meet EEPS targets (for Xcel, EEPS targets established by CO PUC; and for Aquila, the minumum statutory EEPS targets).
	Medium	Same as Low Case
	High	Same as Low Case through 2010, but IOU savings rises to 1.5% of retail sales by 2020, which is slightly higher than EEPS targets.
CT	Low	IOU spending continues at 2008 budget level, 1.7% of revenues.
	Medium	IOU spending doubles from 2008 budget level, ramping up to 3.4% of revenues by 2013 and remaining at that level through 2020.
	High	IOUs meet state policy of acquiring all cost-effective EE (based on efficiency potential study results reported in CT Energy Excellence Report), subject to an assumed annual spending cap equal to 6% of revenues from retail sales.
DC	Low	Spending continues at 2011 authorized SBC rate (1.5 mills/kWh) through 2020, with an assumed 80% directed towards electric EE (and the remainder towards renewable energy).
	Medium	Savings ramp up from projected level in 2011 under Low Case (0.6% of retail sales) to 0.7% of retail sales in 2020.
	High	The same as the Medium Case, except savings ramp up to 1.2% of retail sales in 2020.
DE	Low	Statewide funding for electric EE remains at $6.5M/yr (ACEEE's estimate of authorized 2009-11 annual funding for the Deleware Sustainable Energy Utility) through 2020.
	Medium	Statewide savings ramp up from 0.3% of retail sales in 2011 (as in Low Case) to 0.7% of retail sales in 2020.
	High	The same as the Medium Case, except that statewide savings ramp up to 1.2% of retail sales in 2020.

State	Case	Scenario Description*
HI	Low	Statewide savings exhaust, but do not exceed, the 50% EE allowance under state RPS.
	Medium	Same as Low Case
	High	Same as Low/Medium Cases through 2011, but savings increases from 1.2% of retail sales in 2011 to 2.0% in 2020.
IA	Low	Statewide funding (IOUs & POUs) continues indefinitely at 2008 approved budget (1.2% of retail sales, based on CEE 2007 electric EE budget data)
	Medium	For 2009-2013, IOU spending projections are based on the proposed budgets in their 2009-2013 EE plans, and POU EE spending is equal to the average annual spending in 2003-2005 (data from IUB presentation). For 2014-2020, continue funding at projected 2013 level (3.3% of revenues).
	High	All utilities (IOUs & POUs) achieve savings equal to 1.5% of retail sales by 2011 (the target that IOUs were required to evaluate in their most recent EE plans) and continue at that level thereafter.
ID	Low	IOU savings through 2020 are based on projections from each utility's IRP (Idaho Power's 2008 IRP Update, Avista's 2007 IRP, and PacifiCorp's 2007 IRP update), which decline in aggregate from 0.8% of retail sales in 2010 to 0.4% in 2020; POU savings in each year are 0.25% of retail sales lower than IOU savings.
	Medium	IOU savings through 2010 are based on projections from each utility's IRP, and remain constant at 0.8% of retail sales from 2010-2020; POU savings in each year are 0.25% of retail sales less than IOU savings.
	High	Same as Medium Case, but IOU savings ramp up to 1.5% of retail sales in 2020.
IL	Low	Savings through 2010 are based on IOU 2008-10 EE plans. From 2011-2020, savings equal EEPS targets until statutory cost caps are reached, which is projected to occur in 2012, when savings equal 0.9% of retail sales.
	Medium	Savings through 2010 are based on IOU EE plans. Statutory EEPS cost cap is assumed to be lifted or increased, and savings rise to 1.3% of retail sales in 2013, but remain at that level through 2020.
	High	Savings through 2010 are based on IOU EE plans. Statutory EEPS cost cap is assumed to be lifted or increased, and EEPS targets are assumed to be fully achieved.
MA	Low	IOU savings ramp up to 1.5% of retail electricity sales in 2020.
	Medium	IOU savings ramp up to 2.2% of retail sales in 2012, as proposed in 2010-2012 statewide electric EE plan (April 2009), and remain at that level through 2020.
	High	IOUs meet state policy of acquiring all cost-effective EE, with savings estimated based on the EE potential study conducted for CT, subject to an assumed annual spending cap equal to 8% of revenue from retail sales.

Table A-5. (Continued)

State	Case	Scenario Description*
MD	Low	IOU and SMECO spending through 2013 based on their approved 2008 EE plans, but with a 2-year lag. For 2014-2020, spending remains flat at 2013 level, equal to 1.7% of revenues.
	Medium	IOU and SMECO spending through 2016 based on their approved 2008 EE plans, but with a 1-year lag. For 2017-2020, spending remains flat at the 2016 level, equal to 2.1% of revenues.
	High	IOU and SMECO spending through 2012 based on their approved 2008 EE plans. For 2012-2015, savings are based on achieving 50% of EmPower MD goals through utility programs. Savings from 2016-2020 remain flat at the 2015 level.
ME	Low	Spending from 2009-2012 based on Efficiency Maine's projected budget, and remains at 2012 level, equal to 1.0% of revenues from retail sales, through 2020.
	Medium	Savings ramp up from 0.6% of retail sales in 2010, as estimated from Efficiency Maine's proposed budget for that year, to 1.2% of retail sales in 2020.
	High	The same as the Medium, except savings ramp up to 2.0% of retail sales in 2020.
MI	Low	Statewide savings based on full compliance with statutory EEPS targets, which reach 1.0% of retail sales in 2012.
	Medium	Same as Low Case.
	High	Assumes that statewide savings ramp up from 1.0% of retail sales in 2012 to 1.2% in 2020.
MN	Low	All utilities meet the minimum 1.0% EEPS target for conservation improvement programs.
	Medium	Savings ramp up from 1.0% of retail sales in 2010 to 1.2% in 2020.
	High	Same as Medium Case, except savings ramp up to 2.0% in 2020.
MT	Low	NorthWestern savings through 2020 based on the projection identified in its 2007 Electric Supply Resource Procurement Plan, which remains flat at approximately 0.5% of retail sales per year.
	Medium	Same as Low Case
	High	NorthWestern savings based on 2007 Procurement Plan projection through 2010, but then ramps up from 0.5% of retail sales in 2020 to 1.5% in 2020.
NC	Low	IOUs exhaust, but do not exceed, the EE allowance under the state RPS; POUs first exhaust their allowable use of large hydro, and then meet 75% of remaining RPS needs (after all set-asides are met) with EE.
	Medium	Same as Low Case
	High	All utilities ramp up from annual savings levels of 0.2%of retail sales in 2010 (as projected for that year in the Low/Medium Cases) to 1.0% in 2020.

State	Case	Scenario Description*
NH	Low	Spending levels through 2020 are based on a continuation of current SBC funding levels for electric EE, equal to 1.2% of revenues from statewide electricity retail sales.
	Medium	Savings ramp up from 0.6% of retail sales in 2010 (the projected savings under current funding levels) to 1.2% of retail sales in 2020.
	High	Same as the Medium Case, except savings ramp up to 2.0% of retail sales in 2020.
NJ	Low	2009-2012 spending based on NJCEP funding levels approved by NJBPU; funding from 2013-2020, a percentage of revenues, remains constant at the approved 2012 level.
	Medium	Equal to mid-point between Low and High case projections
	High	2009-2012 spending same as Low Case. From 2013-2020, annual savings are equal 1,500 GWh/yr, the average level needed to meet the 2020 savings goal in the state's draft Energy Master Plan.
NM	Low	IOUs and municipal utilities acheive full compliance with EEPS targets, which requires annual savings of approximately 0.5% of retail sales through 2014, and 0.7% of retail sales from 2015-2020.
	Medium	Same as Low Case
	High	Same as Low/Medium Cases through 2010; but savings then ramp up to 1.5% of retail sales in 2020
NV	Low	IOU savings through 2015 exhaust, but do not exceed, the maximum EE allowance under the RPS. Savings from 2016-2020 remain flat at 2015 level (0.7% of retail sales), rather than tapering off, as would be the case if savings were equal only to the RPS EE allowance in those years.
	Medium	2008-2010 savings equal 0.8% of retail sales, as indicated in Nevada Power's 2008-10 DSM plan, then rise to 1.0% in 2015 and remain at that level through 2020
	High	Same as Medium Case through 2010, but then savings rise to 1.5% in 2020.
NY	Low	Spending from 2010-2020 remains constant at 1.5% of revenues; for comparison, 2008 budget was 1.1% of revenues
	Medium	Savings from 2009-2015 are based on the portion of the state's overall EEPS target allocated by the NYPSC to ratepayer-funded programs. Spending from 2016-2020 remains constant at the average level during 2009-2015, equal to 2.8% of revenues.
	High	Same as Medium Case through 2015. Spending from 2016-2020 remains constant at 3.8% of revenues.
OH	Low	IOU savings equal to 50% of legislated EEPS targets, rising to 1.0% of retail sales by 2019; stipulated savings reflect high assumed levels of mercantile customer opt-out, reliance on T&D measures, and non-compliance.
	Medium	Same as Low Case

Table A-5. (Continued)

State	Case	Scenario Description*
	High	IOU savings equal to 75% of legislated EEPS targets, rising to 1.5% of retail sales by 2019; stipulated savings reflects lower assumed levels (relative to Low/Medium Case) of mercantile customer opt-out, reliance on T&D measures, and non- compliance.
OR	Low	IOU savings through 2020 are based on projections from each utility's IRP (PGE's 2007 IRP, PacifiCorp's 2007 IRP update, and Idaho Power's 2008 IRP Update); POU savings in each year are 0.25% of retail sales lower than IOU savings.
	Medium	Same as Low Case
	High	Same as Low/Medium Cases through 2010, but IOU savings ramp up to 1.5% of retail sales in 2020.
PA	Low	IOU savings meet 2013 target and remain constant through 2020 at the average savings level achieved during 2010-2013, equal to 0.7% of retail sales.
	Medium	Same as Low Case
	High	Same as Low/Medium case through 2013, but then IOU savings increase to 1.2% of retail sales in 2020
RI	Low	Spending through 2020 remains flat at the 2009 level proposed by National Grid in its 2009 DSM Plan.
	Medium	Based on Medium Case scenario assumptions for MA IOUs
	High	Based on High Case scenario assumptions for MA IOUs
TX	Low	IOU savings meet current EEPS goal (20% of incremental peak demand); no POU savings included in scenario.
	Medium	Statewide savings (IOUs & POUs) meet current EEPS target through 2014, and higher proposed target of 30% of incremental peak demand in 2015-2020.
	High	Statewide savings (IOUs & POUs) increases from 0.1% of retail sales in 2008 to 1.0% in 2020.
UT	Low	Savings through 2020 based on PacifiCorp's 2007 IRP update.
	Medium	Savings through 2010 based on PacifiCorp's 2007 IRP update, then rise to 1.0% of retail sales in 2015 and remain at that Level
	High	Same as Medium Case, except savings rise to 1.5% of retail sales in 2020
VA	Low	Statewide savings ramp up from 0% of retail sales in 2010 to 0.3% in 2020.
	Medium	Statewide savings ramp up from 0% of retail sales in 2010 to 0.5% in 2020.
	High	The proposed savings goals are formally adopted and achieved, with annual savings reaching 1.0% of retail sales in 2013, and remaining constant at that level through 2020.
VT	Low	Spending from 2009-2011 is based on Efficiency Vermont's approved 2009-2011 budget plus Burlington Electric's 2008 EE

State	Case	Scenario Description*
		budget; spending from 2012-2020 remains constant at 2011 level, equal to 7.2% of revenues.
	Medium	Same as Low Case
	High	Savings ramp up from 2.1% of retail sales in 2011 (the projected level under the Low/Medium Cases) to 2.5% in 2020.
WA	Low	Statewide savings are based on meeting state goal of acquiring of all achievable cost-effective potential; magnitude of potential based on NWPCC potential study results and an assumed avoided cost of $45/MWh.
	Medium	Same as Low Case, except savings are based on acquiring cost-effective potential at an avoided cost of $85/MWh
	High	Savings ramp up from 1.0% of retail sales in 2010 (the level achieved in 2010-2020 under the Medium Case) to 1.5% in 2020.
WI	Low	Spending through 2020 remains constant at 2008 level of 1.2% of revenues
	Medium	Spending doubles to 2.4% of revenues over 2009-2011
	High	Savings based on achieving EEPS policy recommendation in Governor's Task Force on Global Warming.

* Where scenarios are defined in terms of assumed savings, spending levels are derived from savings. Also, unless otherwise indicated, spending and savings projections for municipal utilities and cooperatives are developed using Tier II approach.

Tier II Scenario Definitions

For Tier II states (and for municipal utilities and cooperatives in many Tier I states), spending projections were developed by employing a standardized set of assumptions about annual spending on ratepayer-funded energy efficiency, as a percentage of revenues (see Table 6). In the low-case scenario, ratepayer-funding for electric energy efficiency programs increases linearly from 0.1% of revenues in 2008 to 0.3% by 2020. In the medium case, spending increases from 0.1% of revenues in 2008 to 0.5% by 2012, and remains at that level through 2020. In the high case, spending increases linearly from 0.1% of revenues in 2008 to 0.8% by 2020 (which is slightly above the current national average of ~0.7% of revenues). The spending assumptions for Florida differ slightly from the other Tier II states, because Florida utilities currently administer electric energy efficiency programs with higher spending levels (i.e., ~0.4% of revenues) than other Tier II states. In the low case, Florida spending remains at 0.4% of revenues through 2020. The medium and high cases for Florida are the same as for the other Tier II states, except that the initial spending level in 2008 is 0.4% of revenues (rather than 0.1%).

Table A-6. Tier II State Spending Scenarios

Case		Ratepayer-Funded Electric EE Spending (% of Revenues from Retail Electricity Sales)												
		2008	2009	2010	2011	2012	2013	2014	2015	2016	2017	2018	2019	2020
Generic	Low	0.1%	0.1%	0.1%	0.2%	0.2%	0.2%	0.2%	0.2%	0.2%	0.3%	0.3%	0.3%	0.3%
Generic	Medium	0.1%	0.2%	0.3%	0.4%	0.5%	0.5%	0.5%	0.5%	0.5%	0.5%	0.5%	0.5%	0.5%
Generic	High	0.1%	0.2%	0.3%	0.4%	0.5%	0.5%	0.6%	0.6%	0.7%	0.7%	0.7%	0.8%	0.8%
FL	Low	0.4%	0.4%	0.4%	0.4%	0.4%	0.4%	0.4%	0.4%	0.4%	0.4%	0.4%	0.4%	0.4%
FL	Medium	0.4%	0.4%	0.4%	0.4%	0.5%	0.5%	0.5%	0.5%	0.5%	0.5%	0.5%	0.5%	0.5%
FL	High	0.4%	0.4%	0.4%	0.4%	0.5%	0.5%	0.6%	0.6%	0.7%	0.7%	0.7%	0.8%	0.8%

A1.3. Cost of Savings Assumptions

Depending on the particular state and scenario, the spending projection may have been estimated from projected first-year savings, or vice-versa. In either case, first-year savings were translated into annual spending (or vice-versa) using an assumed cost of savings. We assume that the average cost of savings depends on the savings level achieved. To capture this relationship, we developed a generic "cost function" that relates the average cost of first-year electricity savings to the savings level expressed as a percentage of the utility (or state)'s retail sales (see Figure 1). The y-axis values in the figure are expressed on a normalized (unit-less) basis, with a cost of 1.00 at a savings level equal to 1.0% of retail sales. The rationale for this cost function is to reflect the fact that, based on our review of energy efficiency program experience, utility costs to acquire savings (on a dollar-per-MWh basis) can be somewhat higher when portfolio savings levels are low (i.e., annual savings <0.5% of retail sales), due to the effect of fixed program delivery costs and because the utility is implementing pilot programs or is ramping up the administrative and delivery infrastructure. There is also evidence to suggest that program costs increase at relatively high savings targets (i.e., annual savings >1.4% of retail sales) either because rebate levels may increase in order to achieve higher market penetration or because the utility includes more expensive energy efficiency measures in its program portfolio. The cost function was then applied to each state by "scaling" the generic cost function based on either state-specific program cost data or an assumed average cost of savings at savings equal to 1.0% of retail sales.[18] Average program costs in each year were escalated for inflation, using the AEO2009 reference case forecast of the GDP chain-type price index.

Table 7 describes the state-specific program cost data, which are derived from recent program results or recently-approved program plans. All states not listed in Table 7 were categorized as either a Low-Cost state or a High-Cost state.[19] Low-Cost states were assumed to have average program costs equal to the national average in 2006: $200 per 1st-yr. MWh saved at a savings level of 0.2% of retail sales, derived from data compiled by ACEEE (Eldridge et al. 2008). High-Cost states were assumed to have average program costs equal to $275 per 1st-yr. MWh saved at a savings level of 1.0% of retail sales, which is based roughly on average costs currently observed among some Northeastern states.

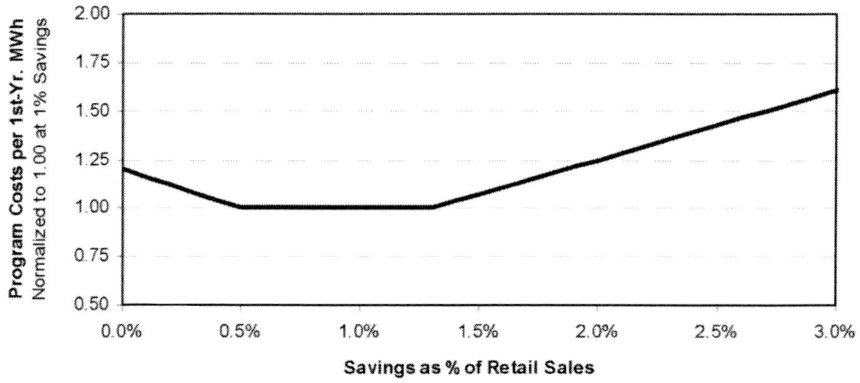

Figure A-1. Generic Program Cost Function.

Table A-7. State Specific Program Cost Data

State	Unit Cost (2007$ per 1st-yr MWh Saved)	Savings Level (% of retail sales)	Data Source and Notes
CA	246	0.8%	Unit cost value is derived from SCE verified savings data and budget for program years 2006 /2007. Savings level value is derived from PG&E, SCE, and SDG&E verified savings in program years 2006/2007.
CT	276	1.1%	Based on the IOUs' 2007 C&LM program results.
IA	166	1.1%	Based estimated savings and spending in MidAmerican's and IP&L's 2009-13 EE program plans
IL	192	0.7%	Based on estimated savings and spending in 2010 from the IOUs' 2008-2010 EE program plans, including programs administered by the DCEO
MA	266	0.9%	Based on the IOUs' 2005 EE program annual reports
MD	177	0.8%	Based on the proposed spending and savings in 2009-2011 from the IOUs' and SMECO's 2009-2015 EE program plans
NJ	168	0.7%	Based on proposed 2008 spending and savings for the NJCEP
NY	284	0.9%	Estimated from data cited in the NYPSC order adopting annual savings targets (6/23/08). The

State	Unit Cost (2007$ per 1st-yr MWh Saved)	Savings Level (% of retail sales)	Data Source and Notes
			PSC order indicates that the incremental annual EE savings above current annual savings levels will cost $305/MWh and that this is 25% higher than current program costs. Given the size of the proposed savings levels relative to current annual savings, this implies an average cost of savings of $284/MWh.
TX	188	0.2%	Based on the IOUs' 2007 program results
WA	260	1.0%	Estimated from market potential and total resource cost data provided by Tom Eckman (NPCC). We estimate program spending from the TRC by assuming that programs cover 50% of the incremental measure cost, and non-incentive program costs represent 25% of total program spending.
WI	213	0.5%	Based on 2006 statewide spending and savings data reported by ACEEE.

A2. Natural Gas Energy Efficiency Spending Projections

Low, medium, and high projections of spending on ratepayer-funded natural gas energy efficiency programs through 2020 were also developed. Given that spending on natural gas programs represents a relatively small portion of total (electric plus gas) ratepayer-funded spending, we utilized a simpler, and more standardized approach to project future spending, compared to the electric energy efficiency projections.

Table A-8. Segmentation of States into Group I and Group II for Natural Gas Projections

Tier I	CA, CT, FL, IA, MA, ME, MN, NH, NJ, OR, RI, UT, VT, WA, WI
Tier II	AK, AL, AR, AZ, CO, DC, DE, GA, HI, ID, IL, IN, KS, KY, LA, MD, MI, MO, MS, MT, NC, ND, NE, NM, NV, NY, OH, OK, PA, SC, SD, TN, TX, VA, WV, WY

A2.1. Revenue Projections

Projections of revenue from retail natural gas sales to residential, commercial, and industrial customers (i.e., excluding sales to electric utilities) were developed in a similar manner as the baseline projections of revenue from retail electricity sales. Retail sales and retail price projections were first developed for each state by applying annual growth rate projections from the AEO2009 reference case forecast (April 2009 update) to actual 2007 retail sales and price data for each state, as reported by EIA. Retail gas sales include sales to residential, commercial, industrial, and transportation sectors, but exclude sales to the electric power sector. Average annual retail gas prices were calculated as the average of EIA's forecast of prices for the residential, commercial, industrial, and transportation sectors, weighted by the quantity of sales to each sector. The natural gas retail sales and retail price projections in AEO2009 are specified at the census-region level. Thus, the census-level growth rates were applied to each state in the respective region. Revenue projections were calculated by multiplying projected retail gas prices by projected retail gas sales, and were converted to nominal dollars using the AEO2009 reference case forecast of the GDP chain-type price index. Unlike the electricity revenue projections, no adjustments were made to the natural gas revenue projections to account for differing levels of energy efficiency savings across scenarios.

A2.2. Scenario Definitions

States were categorized into one of two groups. Tier I states are those with 2008 natural gas efficiency budgets greater than 0.3% of revenues (approximately the national average in that year), based on CEE 2008 budget data. Tier II consists of states with funding below that level and for which 2008 budget data was unavailable. Table 8 identifies which states are in each group.

For most Tier I states, spending scenarios for gas energy efficiency programs were developed by simply stipulating the increase in spending, as a percentage of revenues, from 2008 to 2020, and assuming a linear ramp up to those spending levels. The stipulated spending increases, relative to 2008, are 0% of revenues for the low case, 0.2% for the medium case, and 0.4% for the medium case. There were several exceptions to this standardized approach, made for the three states with the largest natural gas efficiency budgets in 2008:

- For California, the low and medium scenarios assume statewide spending in each year is equal to ratio of the CPUC's long-term gas savings goal in that year relative to the 2008 goal, multiplied by the statewide 2008 natural gas efficiency budget (as reported by CEE). Because the CPUC's long-term natural gas savings goals decline over time, so do the projected spending levels in the low and medium cases. In the high case, we assume that spending on natural gas efficiency remains constant at the level of the 2008 budget, as a percentage of revenues.
- For New Jersey, in the high case, we assume that natural gas spending for 2009-2012 is equal to the approved budget for those years, and remains constant at the 2012 dollar amount through 2020. The low and medium cases are based on the standardized approach for Group I states.
- For Wisconsin, in the high case, we assume that natural gas spending through 2020 is equal to the level indicated in the Governor's Task Force on Global Warming: Interim Report (Feb 2008) as required to meet the proposed natural gas EEPS. The low and medium cases are based on the standardized approach for Group I states.

For Tier II states, gas efficiency spending levels were projected by stipulating the 2020 savings, as a percentage of revenue, for each scenario, and assuming a linear ramp-up from current spending levels. The stipulated 2020 spending levels were: in the low case, the greater of 0% of gas utility revenues or current spending; in the medium case, 0.3% of gas utility revenues; and in the high case, 0.5% of gas utility revenues.

References

APS (Arizona Public Service Company). (2009). *Resource Plan Report.* January 29, 2009.

CEE (Consortium for Energy Efficiency). (2008). *2008 Annual Industry Report.* http://www.cee1.org/ee-pe/2008.

COPUC (Colorado Public Utilities Commission). (2008). Decision Number R08-1243. Docket Number 08A-366EG. December 18, 2008.

CPUC (California Public Utilities Commission). (2008). *California's Long-Term Energy Efficiency Strategic Plan.* http://www. Californiaenergy efficiency.com/index.shtml.

Efficiency Vermont. (2009). *Efficiency Vermont Preliminary Annual Report.* March 2009

Eldridge, Maggie, Max Neubauer, Dan York, Shruti Vaidyanathan, Anna Chittum, and Steven Nadel. (2008). *The 2008 State Energy Efficiency Scorecard.* Washington, D.C.: American Council for an Energy Efficient Economy. http://www.aceee.org/pubs/e086.htm.

EPA (Environmental Protection Agency). (2009). *EPA Analysis of the American Clean Energy and Security Act of 2009, H.R. 2554 in the 111th Congress.* http://www.epa.gov/climatechange/economics/economicanaly ses.html#wax.

Goldman, Charles, Merrian Fuller, Jane Peters, Marjorie McRae, Susan Lutzenhiser & Matt Spahic. (2009). *Energy Efficiency Services Sector: Workforce Size and Expectations for Growth.* Draft of forthcoming report. Berkeley, Calif.: Lawrence Berkeley National Laboratory.

Hopper, Nicole, Galen Barbose, Charles Goldman & Jeff Schlegel. (2008). "Energy Efficiency as a Preferred Resource: Evidence from Utility Resource Plans in the Western United States and Canada." *Energy Efficiency*, 14(2):126-34, October 2008. http://www. springerlink.com/ content/6k46076226625763.

ICC (Illinois Commerce Commission). (2008a). Order issued in Docket Number 07-0539. February 6, 2008.

ICC. (2008b). Order issued in Docket Number 07-0540. February 6, 2008.

Loper, Joe, Jeff Harris, Lowell Ungar, Steve Capanna & Selin Devranoglu. (2008). *Deal or No Deal? Pros and Cons of Trading Under an Energy Efficiency Resource Standard.* Proceedings of 2008 Summer Study on Energy Efficiency in Buildings. Washington, D.C.: American Council for an Energy Efficient Economy.

MDPSC (Maryland Public Service Commission). (2008a). Order 82383. Case Number 9153. December 31, 2008.

MDPSC (Maryland Public Service Commission). (2008b). Order 82384. Case Number 9154. December 31, 2008.

MDPSC (Maryland Public Service Commission). (2008c). Order 82385. Case Number 9155. December 31, 2008.

MDPSC (Maryland Public Service Commission). (2008d). Order 82386. Case Number 9156. December 31, 2008.

MDPSC (Maryland Public Service Commission). (2008e). Order 82387. Case Number 9157. December 31, 2008.

MEEAC (Massachusetts Energy Efficiency Advisory Council). (2009). *The Context for Energy Efficiency Savings: Electric Savings Update, Electric Net Benefits and Gas Savings*. Presentation by Council Consultants at MEEAC meeting. March 24, 2009. http://www.ma- eeac.org.

NJBPU (New Jersey Board of Public Utilities). (2008). *Order Establishing 2009 -2012 Funding Level*. Docket Number EQO7030203. September 30, 2008.

NPC (Nevada Power Company). (2006). *2006 Integrated Resource Plan*. June 30, 2006.

NYPSC (New York Public Service Commission). (2008). *Order Establishing Energy Efficiency Portfolio Standard and Approving Programs*. Case 07-M-0548. June 23, 2008.

SPPC (Sierra Pacific Power Company). (2007). *2007 Integrated Resource Plan*. June 29, 2007.

U.S. DOE (United States Department of Energy). (2009). *Financial Assistance Funding Opportunity Notice: State Energy Program Formula Grants*. March 12, 2009

End Notes

[1] This figure excludes load management.

[2] Depending on the state, ratepayer-funded energy efficiency programs may be administered by utilities, state agencies, or non-profit organizations.

[3] Not evident in the table are the various small states that have relatively aggressive energy efficiency budgets in proportion to their population and retail electricity and gas sales (e.g., VT, OR, ID, IA, RI, NH, UT).

[4] Note that these data represent *approved budgets*, not actual spending.

[5] The state's market transformation goals include achieving zero net energy in all new residential construction by 2020, in all new commercial construction by 2030, and in 50% of existing commercial buildings by 2030.

[6] The lower savings potential is partially a result of the high saturation of many energy efficiency measures and California's aggressive state building codes and appliance efficiency standards; which also reduces the remaining energy efficiency potential that can be obtained through utility programs. The updated market potential study does not, however, account for emerging technologies that may replenish the reservoir of cost-effective savings opportunities over the study period.

[7] Utilities in several of these states (MD, IL, MI) offered energy efficiency programs in the late 1980s and early 1990s, often driven by integrated resource plans, but these energy efficiency efforts were typically abandoned with the onset of electricity restructuring.

[8] A more-complete description of the methodology used to develop the spending projections is available at http://eetd.lbl.gov/EA/EMP/ee-pubs.html.

9 For example, Illinois and Ohio both have EEPS targets that rise to 2% of retail sales per year, which exceeds the level currently being achieved in virtually all leading states. Moreover, Illinois' EEPS includes a cost cap that, unless modified, is likely to preclude attainment of the statutory savings targets in later years. Ohio's EEPS contains provisions to allow many commercial and industrial customers (referred to as "mercantile customers" in the EEPS statute) to opt out of making funding contributions to the utilities' energy efficiency programs if they can demonstrate investments in energy efficiency at their facility. In Maryland, the EEPS is a voluntary goal, and may be achieved through a combination of ratepayer-funded energy efficiency programs, codes & standards, and other types of market interventions. The Virginia EEPS has not yet been passed into law (as of June 2009), and is assumed to be achieved only in the high case.

10 By "incremental annual energy savings" we refer to the annual energy savings resulting from new energy efficiency measures installed in a given year. We estimated 2008 electric energy efficiency savings by applying the savings-to-spending ratio for 2006 (based on the most recent national savings data published ACEEE) to the 2008 electric EE budget (from CEE). For ACEEE data, see Eldridge et al. (2008).

11 For a further explanation of this argument, see Loper et al. (2008).

12 The cap on covered greenhouse gas emissions begins at 3% below 2005 levels in 2012, with the percentage reduction relative to 2005 levels increasing each calendar year, reaching 83% below 2005 levels by 2050. The 2020 target is 17% below 2005 levels.

13 The 2020 GHG emission target for covered sectors is equal to 5,056 mmtCO_2e, compared to EPA's reference case projection of 5968 mmtCO_2e for covered sectors in 2020 (EPA 2009).

14 This range in emission reductions was calculated by reducing our projections of electricity savings from ratepayer-funded programs by the portion assumed to be implicit in the EPA reference forecast, yielding a range in cumulative savings over 2009-2020 equal to 116-270 billion kWh. We then applied to that range of savings emission rates ranging from 0.36-0.86 metric tons per MWh. The emission rate range reflects uncertainty and geographical diversity in the emission rate of generation offset by future energy efficiency, with the lower end of the range equal to the emission rate of a combined cycle gas turbine (CCGT) generator and the upper end equal to the mid-point between the emission rates of CCGT and sub-critical pulverized coal generation.

15 The economic recession has resulted in a significant reduction in sales of many electric utilities, in some cases resulting in negative growth rates. In this environment, some utilities are concerned about under-recovery of fixed costs (and authorized earnings) because a significant portion of their revenues are derived from volumetric-based rates. Significant increases in funding for energy efficiency may exacerbate this problem.

16 A few states (e.g. PA) have already decided that program administrators can take full credit for savings if customers receive a rebate from the utility, even if customers access additional funding sources.

17 Products for which new standards were adopted in 2007 Energy Bill include: general service lamps, reflector lamps, residential boilers, clothes washers, dishwashers, dehumidifiers, electric motors, metal halide lamp fixtures used in high-ceiling commercial and industrial applications, walk-in coolers and freezers, and external power supplies. See Appliance Standards Awareness Project factsheet at http://www.standardsasap.org/documents/2007EnergyBill_Standardsfactsheet.pdf

18 For example, if data for a given state indicate that average program costs are $200 per 1st-yr. MWh saved at savings equal to 1.0% of retail sales, then the generic cost function would yield an average cost of $250 per 1st-yr. MWh at savings equal to 2.0% of retail sales (i.e., 1.25 times the cost at a savings level equal to 1.0% of retail sales).

19 High-Cost states consist of: CO, HI, NH, and VT. All others (not listed in Table A-7) are deemed Low-Cost.

INDEX

U

W

Y